Mythic Galveston

CREATING THE NORTH AMERICAN LANDSCAPE

Gregory Conniff
Edward K. Muller
David Schuyler
Consulting Editors

George F. Thompson
Series Founder and Director

Published in cooperation with the Center for American Places,
Santa Fe, New Mexico, and Harrisonburg, Virginia

Mythic Galveston
REINVENTING AMERICA'S THIRD COAST

SUSAN WILEY HARDWICK

The Johns Hopkins University Press
Baltimore and London

© 2002 The Johns Hopkins University Press
All rights reserved. Published 2002
Printed in the United States of America on acid-free paper
9 8 7 6 5 4 3 2 1

The Johns Hopkins University Press
2715 North Charles Street
Baltimore, Maryland 21218-4363
www.press.jhu.edu

LIBRARY OF CONGRESS CATALOGING-IN-PUBLICATION DATA

Hardwick, Susan Wiley.
Mythic Galveston : reinventing America's third coast /
Susan Wiley Hardwick.
p. cm. — (Creating the North American landscape)
Includes bibliographical references (p.) and index.
ISBN 0-8018-6887-4 (hardcover : alk. paper)
1. Galveston (Tex.) — History. 2. Galveston (Tex.) — Social conditions.
3. Galveston (Tex.) — Ethnic relations. 4. Human geography — Texas —
Galveston. 5. Regionalism — Texas — History. I. Title. II. Series.
F394.G2 H25 2002
976.4'139 — dc21
2001003755

A catalog record for this book is available from the British Library.

To my first writing teacher,

𝓜ILDRED 𝓕ISHER 𝓦ILEY,

whose enthusiasm for learning laid the foundation.

This one is for you.

Contents

Preface   ix

Acknowledgments   xi

CHAPTER 1
Introduction   1

CHAPTER 2
Early Visitors and Settlers   15

CHAPTER 3
Birth of a Global City   28

CHAPTER 4
Rise of the Elite   59

CHAPTER 5
After the Storm: Reinventing an Island Environment   92

CHAPTER 6
The End of Immigration: The Galveston Movement   119

CHAPTER 7
The Reinvention of People and Place   138

Bibliography   155

Index   169

Preface

How best to analyze the evolution of a place as complex as Galveston? None of my prior research and writing challenges had been this difficult. The analysis of the peopling of this island city made the process even more complex. As this book illustrates, Galveston is truly a place set apart.

Many scholarly and popular works have been published about this fascinating island city in the century since the Great Storm, the worst natural disaster ever to strike North America. Like *Mythic Galveston*, several of these earlier publications presented information about the flow of immigrants—especially German immigrants—who arrived by the tens of thousands at the busy Galveston wharf. None, however, sought to explain the comparative relationship between the settlement histories and social processes of these Texas newcomers.

Prior work has said even less about the settlement patterns and experiences of African American residents of the island, both before and after the Civil War. Further, there has been an almost complete disregard for the sharp boundaries that existed—and continue to exist—between social, ethnic, and racial groups on the island. Even in the early twenty-first century, "Born on the Island" remained an expression whose deeper meaning almost every local resident understood. *Mythic Galveston* addresses this gap in the literature. It focuses on spatial and social relationships—those between various groups of people, and between people and place.

After almost three years spent combing through census records, conducting interviews, sorting through Web sites and their related links on the Internet, squinting into microfilm machines in library corners, tabulating city directory listings, and compiling maps, I find that the real Galveston in many ways remains a mystery. Although many of the questions posed by scholars interested in urban evolution are answered in this book, other questions linger. Some remain as compelling to me at the end of this research project as they did at the beginning. Fascination with Galveston has me in its grip.

Exactly what social and cultural processes have shaped Galveston Island since its earliest settlement by aboriginal people? How did these processes create the island city's distinctive spatial patterns and contribute to its distinctive sense of place? What role did divisions between various groups of residents in the city play in the development and maintenance of Galveston's socioeconomic and political systems? These and other questions provided a structure for the myriad data gathered for this project and established a framework for analysis.

In recent decades, *diversity* has become a buzzword. Galveston Island is perhaps the ultimate expression of this concept. It harbors diverse religious, ethnic, and racial groups and socioeconomic classes, and it is the focus of interrelated environmental concerns. As in many other places, this diversity at times has created conflict as well as potential for growth. After many decades of struggle, however, the city of Galveston is once again alive with potential. But very little of this cultural and economic potential has much to do with the present; rather, most is rooted in the city's colorful and dramatic past.

Over the course of more than a century and a half, Galveston emerged as a place unto itself. Quintessentially a Texas city and always absolutely a southern place, the city has also always been connected beyond its own state and region, to a much larger realm. This book suggests that Galveston and a few other Gulf Coast cities located along a sandy outer ribbon on the extreme edge of a place recently dubbed "America's Third Coast" may indeed form a distinct cultural region, extending as far east as Pensacola.

Mythic Galveston argues that this unique place manifested the process of globalization even in its earliest years of settlement. Regular contact with the Caribbean and points beyond made Galveston a cosmopolitan city, rich in Latin American, European, and African cultural and economic influences. As the photographs and maps in the book demonstrate, Galveston's diverse and still quite international landscape lingers today.

Acknowledgments

No book of this scope could have been written without the help of a sustaining community of support. Without the encouragement and support of "Mr. Ethnic Texas," Professor Terry Jordan-Bychkov, this book would not have been written. As research mentors, colleagues, and friends, Terry and his wife Bella offered support all along the way.

A cadre of helpful graduate students at Southwest Texas State University over the course of the past three years listened patiently while a new Texan asked questions about a place most had visited many times. To my closest research associate, Susan Hume, goes my deepest appreciation for all her countless trips between the library and my office, endless hours searching through census records and city directory listings, and meticulous editing and formatting of the final manuscript. With the help of Lydia Bean, Susan offered ideas prior to the development of each section of the manuscript, read each chapter, made suggestions for revisions, and edited the complete draft several times, a task usually reserved for editors. Without Susan and Lydia's help, patience with my absent-minded obsession about the topic, and warm and accepting friendship, the completion of the final manuscript would not have been such an enjoyable process.

In the first two years of the project, essential research assistance also was provided by graduate student Rebecca Sheehan and two other former graduate students in our department—Joy Adams, now a Ph.D. student working with Terry Jordan at the University of Texas, and Kathy Alexander, an environmentalist at the Texas Natural Resources and Conservation Commission. Their graduate assistant support would have been impossible without the generous assistance of the chair of the Department of Geography at Southwest Texas State, Dr. Lawrence Estaville, and the graduate coordinator, Dr. Fred Shelley. Their continuing interest in my study of Galveston, along with the helpful support provided by other faculty and staff colleagues in the department—especially Allison Moore, Dr. Sally Caldwell, Judy Behrens, Dr. Pam Showalter, Dr. David Stea, Dr. Jim Kimmel, and Dr. Craig Colten—are sincerely appreciated.

Linda Prosperie, head cartographer for the project, created carefully designed Geographic Information Systems–based map analyses that made the comparative analysis of settlement patterns possible in a dramatically new way. I am deeply grateful for Linda's skill, her patience with my ever changing data sets, and her inspiration—despite facing "do or die" doctoral exams one week after the manuscript was submitted. Over the course of the project, her friendship came to mean as much as her help with development of maps and other graphics for the book.

From the earliest days of the research, the energetic and always well-informed ideas and encouragement of archivists and staff members Shelly Henley Kelly and Anna Peebler at the Galveston and Texas History Center at Rosenberg Library in Galveston made all the difference. Their interest in my geographic study was supported by Casey Greene, head of Special Collections at Rosenberg. I am profoundly grateful for Shelly and Anna's willingness to share their vast knowledge of sources on Galveston and for their patience with an outsider who was decidedly not a "BOI." Along with the help of the staff at Rosenberg Library, I appreciated being able to explore and use manuscripts and other archival materials at the Center for American History at the University of Texas at Austin and at the Library of Congress in Washington. Added to these library support services was the invaluable help of interlibrary loan librarians at Alkek Library on my own campus. Their patience and help securing even the most challenging sources proved to be invaluable throughout the research process.

Finally and ultimately, I deeply appreciate the encouragement and final push needed to complete the manuscript offered by George F. Thompson and Randall Jones at the Center for American Places. Without them, the ideas expressed here might have remained forever inside my computer.

To each of these supporters and friends, along with my always supportive husband Donald, and my three patient West Coast sons and their families (who probably never will understand my obsession with the "Third Coast"), I am deeply grateful. Apart from all the help I have been given these past three years, I, of course, remain responsible for any errors or misconceptions that may appear in the following pages.

Mythic Galveston

INTRODUCTION

> Here in Galveston the humidity was like a clammy hand held over your face. Yet the city had a ghostly charm. The scent of the tangled gardens hung heavy on the muggy air. The houses, pock-marked by the salt mist and the sun and heat and mildew, seemed built of ashes. Here was a remnant of haunted beauty—gray, shrouded, crumbling. What did they resemble? Of what did this city remind me? Miss Havisham, of course. That was it. Miss Havisham, the spectral bride in *Great Expectations*.
>
> EDNA FERBER, *A KIND OF MAGIC*

This literary image of Galveston Island captures its timeless grace, its haunting beauty, and also its decaying sense of place. For many Texans, the word *Galveston* brings back memories of sunny days relaxing on the beach or splashing in the warm waves of the Gulf. Others picture more disturbing images of deadly hurricanes, oppressive heat and humidity, and polluted air and water. Whether their perceptions of Galveston Island are extremely positive or depressingly negative, however, few may think of the vital role of this insular city—the largest in the state from 1850 to 1890—as the main port city of the state and much of the midcontinent (map 1.1). Between the mid-1840s and 1917 the wharves at Galveston welcomed tens of thousands of new immigrants from Europe, Africa, Asia, and Latin America. For this reason, Galveston may best be described as the Ellis Island of Texas.

WHY GALVESTON?

The inspiration for this book happened quite suddenly. As a newcomer to Texas, I had taken one of my frequent weekend trips to unknown parts of the state. This trip, however, turned out to be like no other.

My first glimpse of the island, after crossing the long causeway that connects it to the Texas mainland, was both complex and confusing. What

MAP I.I. *Galveston Island site map (Cartography by Linda F. Prosperie)*

did this landscape mean? How did it come to look like this? How could a small city with such wealthy residential neighborhoods also have districts of poverty and low-income housing almost immediately next door? How could an island popular with tourists from many places in the United States also be a place of high-density and often crumbling residential and commercial development? And, why did the architecture in this almost tropical place remind me so much of my home in the Northeast or even of my visits to Germany and other places in Europe? Obviously, I had much to learn.

After parking in the Strand Historic District, I visited the city's gentrified downtown district. As I continued to walk toward the eastern end of the island, I noticed an old customs house (fig. 1.1). As a geographer who had spent more than twenty years studying the effects of immigration in California, I thought for a moment that I was standing by the old customs house in San Francisco. Could this small Texas island, now a popular tourist destination and resort city, have been a port of entry for new U.S. settlers? It appeared so from this old building and from the many historic homes and businesses still standing in this decaying, but still quite lovely, downtown neighborhood.

This first glimpse of Galveston Island brought out the excited "field geographer" in me. Months later, after visiting historical museums and vine-covered cemeteries, and talking at length with old-timers, shopkeepers, and even historians now living on the island, I pondered several unanswered questions: Exactly who came to this place, and why? Where did the newcomers settle after their arrival on this tiny island? Did they stay in Galveston or move on, into the interior of the state and nation? Finally, and perhaps most important, why would anyone choose such an isolated, storm-prone place as the ultimate destination for their long journey into a new life?

Over the next three years I combed through manuscript census records, examined city directories, read the numerous secondary sources published about the island's historical development, and interviewed historians, longterm residents, historic preservationists, entrepreneurs, and investors in island property. I was surprised to discover that nearly 40 percent of Galveston's population was German when the earliest census records were tabulated in 1850. I also learned that the city's cosmopolitan population, oleander-lined streets, and lovely Victorian architecture reflected its late nineteenth-century status as the most international city in Texas.

Today, despite the destruction caused by several hurricanes, and in the face of the economic dominance of nearby Houston, Galveston remains an ethnically rich and culturally vibrant community. Ethnic, racial, and socio-economic diversity abounds. One of the city's claims to fame in the year 2000 was that it had more government housing built for low-income residents than any other city of its size in the country. Practically next door

FIGURE I.I. *My interest in Galveston's ethnic geography began at this customs house near the historic Strand District. (Photo: Susan E. Hume)*

to low-income houses in Galveston are the stately houses of some of the state's, and the nation's, wealthiest citizens (figs. 1.2, 1.3).

From these few but dramatic examples of the diversity of life on Galveston Island, it is clear that Galveston has long been a fascinating and complex place. Numerous historical accounts chronicle the development of the city, which played such a prominent role in the peopling of the state of Texas and areas beyond. Yet until now, no one has documented and analyzed the ethnic patterns and processes of this unique place. Of particular note is the absence of published materials that document and compare the residential patterns of immigrant groups that lived in Galveston over the years.

THE PLACE

Galveston, like Tyre, is built upon an island in the midst of the sea, and if her people, like the Tyrians of old, continue to be economical, industrious, and enterprising; sincere, faithful, and hospitable to strangers . . . she will become the center of commerce, the resort of all nations, and attain the wealth and power, and it may be . . . the greatness and glory of the ancient city.
FREDRICK BENJAMIN PAGE, 1845

A promotional article describing Galveston's potential for developing into a large city of unlimited wealth, power, and influence described the area and its people in glowing terms. Clearly, the earliest perceptions of the island were often positive but were also many times incorrect. We will see that the island's physical environment, along with its limited size and lack of natural resources, made its long-term settlement and economic development extremely difficult, even for the biggest dreamers.

First incorporated as a city in 1839, Galveston is located on the eastern end of a narrow barrier island on the edge of the Texas Gulf Coast. Despite the island's almost complete lack of the natural resources needed to support a dense population, early developers believed that its location as the best natural harbor on the Texas coast would more than outweigh these limitations on development and growth. In the nineteenth century Galveston grew dramatically. Between 1850 and 1890 it was the largest city in Texas—unequaled in cosmopolitanism, beautiful residences, commercial development, and booming industrial activity.

Galveston also bulged with high-energy immigrants who had come to Texas to seek a new life in a new land. These newcomers helped build the city. They also supported its economic activities and developed its diverse

FIGURE 1.2. *Galveston's historic residential neighborhoods feature a variety of modest homes like this one. (Photo: Susan E. Hume)*

FIGURE 1.3. *Ashton Villa on Broadway, home to one of Galveston's earliest and wealtiest families, is booked months in advance for weddings and other social gatherings. (Photo: Lee Miller)*

cultural landscapes. By the time the first census was taken in Texas in 1850, Galveston already had four thousand residents. By the turn of the twentieth century, the city's population was reaching toward forty thousand.

Barrier islands like Galveston are elongated landforms lying parallel to the shoreline. They are usually formed mostly of sand and are separated form the mainland by estuaries and lagoons. This type of island occurs in chains along coastlines. Barrier islands protect the mainland by insulating it from the force of wave action associated with tropical storms. Waves, winds, tides, and floods are major forces that create them. Therefore, barrier islands are by their nature ever changing. In the case of Galveston Island, frequent tropical storms and hurricanes alter its physical geography, while seasonal and cyclic changes in wind and wave patterns also play a role in shaping its geomorphology.

Although barrier islands like Galveston are unstable environments, development of these areas is often an attractive prospect. The desire to remake such environments into more stable places by altering them to conform to mainland standards generally results in the destruction of most of the environmental systems that made them attractive in the first place.

Galveston probably should never have been settled in the first place. The positive perceptions of early North American and European visitors and settlers fostered a strong belief in the tremendous potential of the island as an appropriate place for dense settlements, urban expansion, and economic development. On Galveston Island nature wasn't simply altered, it was brought under complete human control. It was manipulated, altered, and in time reinvented. The image of the island's abundant and highly temperate, even tropical, climate was promulgated and sustained through time by immigrants, despite almost constant reminders of the lack of real evidence for such overly optimistic views. When nature did challenge the misperceptions of early residents, they fought back with ever escalating technological fixes. The culmination of these "fixes" was the decision, made after the Great Storm in 1900, to raise the elevation of the city of Galveston, rather than to relocate on the mainland. The story of the development and evolution of the city of Galveston is, therefore, the story of the victory of false environmental perceptions, anthropomorphic arrogance, and the use of technology to dominate nature.

Typical of the distorted thinking of early immigrants is this young woman's expression of her excitement about relocating to Texas from Germany: "One of my relatives who had gone to Texas . . . had come back on a visit to the Rhine. He told many enthusiastic stories about his new home. I was thrilled when I listened to him describe the charms of that recently

INTRODUCTION

opened paradise, the eternally blue sky, the radiant sun ... tempting tropical fruits, Indians, and wild animals too" (Dielmann 1960, 364).

After forty-four days at sea, another immigrant described in more realistic terms the first glimpse of Galveston Island: "Before us lay the longed for coast, but without the avenues of oleanders that had been described to us. A hurricane had just recently caused terrible destruction, uprooting trees and washing them away. Galveston, for Texas a big city, looked to the Europeans like a set-up of paper toys. The houses stood on posts ready to be moved from one place to another" (Dielmann 1960, 365).

In its natural state, exactly what were the opportunities and constraints for settlement and survival on Galveston Island? How rich were the possibilities for its development? Would it be easy for a large and densely settled urban area to develop and grow here, or would the island's environmental limitations impede settlement and subsequent economic, cultural, and sociopolitical development? The answers to these and other related questions frame my discussions.

DATA SOURCES AND METHODS

A combination of quantitative and qualitative methods was used to analyze the data for this book. Especially important was information gleaned from the U.S. Census of Population Manuscript Census for the years 1850–1920. Statistical data on place of birth, age, gender, nativity, family structure, and employment for each of the census years were tabulated and then mapped, made into tables, or used in the narrative of this book. Names and residential addresses of individual immigrants and foreign-born heads of households, in particular, were entered into a database and spatially analyzed. In cases where no addresses were provided by census takers, names were traced in city directory listings to locate individuals' places of residence and employment histories. The use of data from both census records and city directories laid a solid foundation for other methods used to develop the arguments presented here.

Maps showing the results of this analysis also established whether streets and neighborhoods maintained an ethnic identity through time and also provided evidence of the patterns of dispersal or clustering of various ethnic and racial groups. Field observations and photography provided on-site documentation of these mapped patterns.

Archival materials were gathered primarily from the excellent collection on local history housed at the Rosenberg Library in Galveston. Others

TABLE 1.1. **Sample List of Ships Entering Galveston Harbor between the Peak Years of 1846 and 1871**

Name	Date of Arrival	Origin
Andaira Valley	March 27, 1846	Antwerp
Brig Jon Dethard	January 2, 1848	Bremen
Brig Antoinette	June 13, 1848	Bremen
Galliott Flora	May 2, 1849	England
Isabella Teague	May 11, 1849	England
Brig Herschel	May 21, 1849	Bremen
Brig Reform	December 3, 1849	Bremen
Bark Hamburg-Knollen	December 15, 1849	Hamburg
Bark Neptune	June 13, 1850	Bremen
Bark Solan	November 30, 1850	Bremen
Brig Sophie	January 19, 1852	Bremen
Bark E. von Beaulieu	January 1857	Bremen
Brig Anna Louisea	July 31, 1857	Bremen
Wiser	June 1858	Bremen
Bark Iris	October 17, 1866	Germany
Bark Fortuna	November 1, 1867	Bremen
Bark Diana	November 6, 1867	Bremen
Bark Texas	October 19, 1868	Bremen
S.S. Lord Buve	December 9, 1868	Liverpool
Galveston	June 19, 1871	Bremen

Source: Ships' Passenger Lists, Port of Galveston, 1846–71.

were found in the Center for American History at the University of Texas in Austin, in the Texas State Archives, and in historic records from ship's passenger logs kept by the U.S. Immigration and Naturalization Service in Houston and San Francisco, as well as at the Seaport Museum located in Galveston. The ship's manifests of all incoming vessels entering the port of Galveston were analyzed to clarify, document, supplement, and compare the arrival patterns of diverse immigrant settlers of the city, region, and state (table 1.1).

The data set for the years 1846–1920, the peak years of in-migration, was particularly rich and detailed. Two other sources, the Index to Naturalization Records in Texas Courts, 1846–1939, and the Port of Galveston Records, 1846–1950, provided an excellent foundation for the study of Galveston's evolving spatial and social patterns. In addition, church and synagogue records, newspapers, and personal interviews furnished evidence of the city's changing demographic character through time.

Information tabulated from these statistical sources was carefully analyzed and mapped using factor analysis and Geographic Information Systems (GIS) overlays. These statistical and cartographic techniques helped deepen my understanding of the complex settlement patterns of this microstudy and made the data more applicable for macro-analysis of comparable places in other time periods. The key variable in the statistical analysis used throughout this book is ethnic origin. GIS maps for the years 1850, 1880, 1900, and 1902–3 are presented as evidence for the city's unique ethnic and racial residential patterns. A comparative analysis of the maps for each of these key census years, as well as the narrative text that accompanies the maps, is filled with many surprises.

Along with GIS mapping, segregation indexes were used in the final chapter of the book. This comparative statistical method helped identify and map significant spatial clustering of ethnic groups in particular neighborhoods or districts of the city, when the increasing size of the city made it impossible to count and track individual names or even family listings from historic census records.

Data gathered from personal interviews with long-term residents of the island were a key element in interpreting the diverse facts and figures about the island's evolution. Many of these individuals are direct descendants of the earliest German, Russian, and Italian settlers of the city. Information gathered during these interviews helped in the interrogation and corroboration of mapped patterns and also lends human interest to the narrative.

FOCUS OF THIS STUDY

This study is nested within the larger traditions of scholarly and popular work in cultural-historical geography. It builds on geographical and historical data presented in Terry Jordan's seminal book, *German Seed in Texas Soil* (1966), and on his many other publications on the role of this ethnic group in the development of the state. Jordan's recent work of defining a new cultural region on the Gulf Coast—a place he calls the "Creole Coast"—was particularly valuable in framing my research questions and inspiring my final conclusions about Galveston as a significant Gulf Coast pivot city (Jordan-Bychkov 2001).

Research for this study also depended on the work of earlier scholars focusing on the development of the city of Galveston. Recent publications that proved to be most helpful were *The Alleys and Back Buildings of Galveston* (1996) by historic preservationist Ellen Beasley; Elizabeth Turner Hayes's *Women, Culture and Community: Religion and Reform in Galveston, 1880–1920* (1997); *Galveston and the Great West* (1997) by Earle Young; and *Through a Night of Horrors: Voices from the 1900 Great Storm*, edited by Casey Edward Green and Shelly H. Kelly.

The significance of this study lies in its focus on immigrant settlement patterns and the contribution of comparative ethnic and racial groups to the evolving cultural landscape and sense of place in Galveston. A common belief among Texans who live on the island—as well as many who live elsewhere—is that Galveston Island *has always been and still is* one of the most egalitarian places in the state and region. Because of its small size and densely clustered residential neighborhoods, the city is most often perceived as a place where people from all over the world lived in harmony—no matter what the color of their skin, religious beliefs, language, or income. This view paints Galveston as an island paradise where everyone had an equal piece of the economic pie and an equal impact on the cultural landscape. This study debunks that myth.

My initial research for the book was motivated by a fascination with understanding how such a small city in such a confined space could have survived into a new century despite devastating hurricanes, yellow fever epidemics, severe environmental constraints, and extreme social, political, economic, and cultural diversity. But the real significance of Galveston may lie not so much in its survival "despite all odds," but in its relative location. Is Galveston unique among cities in the South, or does it form the western edge of a much larger Gulf Coast region?

My data suggest this island city is a key part of a distinct and as yet undefined geographic region. This coastal belt extends in a narrow strand along the sandy shores of the Gulf Coast, and its offshore barrier islands extend eastward to other small cities, such as Pascagoula, Gulfport, Biloxi, Ocean Springs, Mobile, and Pensacola. Although each of these places is unique, they share a set of characteristics that distinguishes them from other places in regions geographers call the "Gulf Coastal Plain" or even more broadly the "Southeastern Coast."

Jordan's work on the "Creole Coast," and my discussions with him, hint at the existence of this distinctive physiocultural region located along the outer edge of the central Gulf Coast. Physically, settlements in this newly defined region have similar environments grounded in sandy offshore islands

INTRODUCTION

and coastal beaches. All are set along shallow Gulf waters that result from the broad continental shelf. Along most of this coastline, the ocean is very shallow for a considerable distance offshore. All these places have humid, subtropical climates with poorly drained, sandy soils. None has a significant natural resource base to support a large population.

Economically, this new region has long depended on the development of its port and on resources, such as fish, petroleum, and tourism, that relate directly to its maritime orientation. This Gulf Coast region also maintains a distinctive identity because of these unique historical and cultural characteristics:

- Its pre–Civil War Confederate cities experienced the turmoil of war and Reconstruction.
- Racial, ethnic, and cultural diversity is celebrated throughout the region in colorful and often extravagant festivals.
- There is more integration in the residential areas than is found in other southern cities.
- Old-stock immigrant groups have been replaced by more recent arrivals from Vietnam and refugees from Latin American countries, such as Guatemala and El Salvador.
- Growth potential is limited by lack of available space for development and by a lack of resources to support large populations.
- Food and dietary customs differ from those found in the rest of the South (for example, rice, red beans, and wheat bread are favored over typically southern cornbread and chicken-fried steak).
- The cities pivot toward the sea rather than toward the interior, and most depend on exports from their local environment rather than on their hinterland in the interior.
- The cities are increasingly dependent on tourism and other island attractions, such as cruise ships and casinos.
- The cities resonate with Caribbean culture: most have residents from Cuba, Haiti, and elsewhere in the Caribbean, and the resulting visible landscape has features typical of more tropical places.

Exceptions exist within this elongated region because of the unique histories of various cities. New Orleans, for example, has long been closely linked with the interior because of its historic role in the agricultural development of the Mississippi River hinterland. New Orleans, therefore, may not fit this model. Another clear anomaly is Port Arthur, Texas, because of its boom town status, based on the discovery of oil at nearby Spindletop at Beaumont.

CHAPTER SUMMARIES

This book provides a foundation for understanding Galveston's settlement history through time and space. I argue that economic, social, and cultural boundaries between various groups of people in Galveston extended well beyond the color of their skin or their places of worship. Data presented in this study document the existence of a distinct class system based on socioeconomic differences as well as on place of birth. Differences in social class were defined by race, ethnicity, economic status, and always according to the distinction between people who had been born on the island (BOI) and residents who were foreign-born. *Mythic Galveston* discusses these racial, religious, linguistic, and socioeconomic boundaries between people and place, as newcomers struggled to *acculturate in place* to their new environment.

Indeed, every person who settled on the island—including the earliest settlers, the Karankawas—was a newcomer. This makes Galveston a fascinating place to analyze the related themes of migration, settlement, and acculturation. In addition, Galveston's peripheral and isolated geographic location and its prominent role in the settlement of Texas provide an excellent opportunity to study how geographic processes associated with the influx of disparate groups can alter the regional character of space and place.

Chapter 2 presents a brief historical overview of the earliest people to call the island home, the Karankawas. This is followed by a discussion of the comparative historical geography of other pre-1846 groups of Europeans, Latin Americans, and North Americans who came as visitors, explorers, and settlers.

Chapter 3 takes a closer look at the rapidly growing frontier city of Galveston, focusing on its evolution in the period before the Civil War. Comparative maps of foreign-born residential patterns created from data listed in the state's first census year, 1850, provide grounding for the analysis of migration and settlement of people from Belgium, Denmark, Mexico, Poland, Portugal, Spain, Sweden, Wales, Canada, Germany, Switzerland, Scotland, Italy, Ireland, England, and France.

In Chapter 4, data describing Galveston in the second half of the nineteenth century are presented in maps and tables that accompany a narrative analyzing the city's "second wave" of new immigrants. Residential maps of the largest foreign-born immigrant groups listed in the census of 1880 are presented as a central part of this chapter. This information is compared with other maps, which show the locations of Chinese and Italian businesses, and with a map of the African American cultural landscape in the decades just before the Great Storm.

Chapter 5 provides evidence of the sea of changes that swept through Galveston's residential patterns, its politics and economy, and its morphology during and after the Great Storm of 1900. As the settlement patterns of various immigrant groups were rearranged dramatically following the storm, the city took on a completely reorganized form and function. Of special note in this period are the Italians who arrived in Galveston in relatively large numbers after the storm—some to continue their journeys to fertile strawberry fields in Dickinson and others to buy land farther inland. But many stayed in Galveston, adding yet another element to the city's ever intensifying ethnic landscape.

Surprisingly, the in-migration of Italians, Greeks, and other groups continued in the two decades following the storm. Among these new arrivals were over ten thousand people from Russia who were rescued by an international effort organized by Jewish leaders in New York, Galveston, and London. The "Galveston Movement," as this global network was called, is discussed in detail in chapter 6. Most of the Russian-Jewish migrants who made up much of this last large immigration wave in the city's history did not stay in Galveston more than a few days. Nonetheless, their impact on the state and the midcontinent provides yet another example of the significance of the port of Galveston in the peopling of the state and nation.

The ongoing story of the dynamic interaction of people and place in Galveston after 1920 is the subject of chapter 7. In this last census year before the passage of new U.S. laws severely limiting the number of new immigrants allowed to enter the country, Galveston's population had a chance first to stabilize and later to stagnate. In 1920 it appeared that the city's economy would never recover from the cataclysmic events of the preceding two decades of change.

I conclude with a discussion of the city's relatively recent reincarnation as a bed-and-breakfast destination resort for upper-middle-class Texans and other travelers interested in historical architecture and beachcombing. Here again, predictions about Galveston's potentials and the many challenges for its future are presented within the context of a broad-based geographical analysis of space and place.

The Sandusky map (see map 4.1) offers a fitting conclusion for my introductory chapter. Drafted by a Polish immigrant—the first cartographer hired by the city of Galveston—this copy of a lithograph held at the Rosenberg Library in Galveston shows the street layout and overall design of the city as it appeared in 1871. The story told in the following chapters makes it clear that both the physical and human geography of this island city was to change dramatically in the next seventy-five years—a process of change that continues today.

Early Visitors and Settlers

As we drifted onto shore, a wave caught us and heaved the barge, a horseshoe thrown out of the water. The jolt when it hit brought the dead-looking men to. Seeing land at hand, they crawled through the surf to some rocks. Here we made a fire and parched some of our corn. We also found rain water. The men began to regain their senses, their locomotion, and their hopes.
ALVAR NUNEZ CABEZA DE VACA, 6 NOVEMBER 1528

I. Cabeza de Vaca

A gray sky flowed as far as my
eyes could see. Length of shoreline
slipped off into distant horizon, took
me beyond, transported me to clouds flowing
like mud ripples on salt water flats, I
stand alone on oyster shells, the north wind
steady in my face, my past life far away,
poised, yet unable to walk away on the water.

Near the shore's edge, in the late afternoon,
in the dry mudfield, behind the tall grass,
I rest, barefooted, by a fire and a bed of
thatch, transformed by the cold into another.

And the heavens turn silver, and light
breaks through to a vision of wealth, and there
is no sound, but the lapping of waves against the shore:
I, Cabeza de Vaca, will survive.
LAYNE HENDRICK, AUSTIN, TEXAS, 1999

An exotic blend of aboriginal people, Haitians, Louisianans, Mexicans, and European adventurers came together on Galveston Island in its first centuries of settlement. Karankawa Indians and Spanish, French, Dutch, Mexican, English, and North American hunting parties, military men, explorers, pirates, adventurers, and colonists joined the abundant birds, snakes, and plants on the island with its frequent storms. Many arrived by accident. Others came quite suddenly as their ships washed ashore on this unexpected piece of land off the Gulf Coast.

On a windy November day in the early 1960s, an employee of the new Jamaica Beach subdivision on west Galveston Island was surprised to dig up a pile of bones when he was extending the boundaries of the new housing development. The bones were later identified as belonging to the Karankawa Indians, the earliest native settlers on the island. Residing more permanently along the inland coast near Matagorda, these original settlers depended on the rich resources of the island during summer months, for hunting and ceremonial purposes.

GALVESTON'S ABORIGINAL PEOPLE: THE KARANKAWANS

Because the Karankawa people and other groups living along the Texas Gulf Coast and on the offshore islands did not practice agriculture, anthropologists have long referred to the region as a "cultural sink." Unfortunately, the term *sink* has a negative connotation. This all-too-common terminology labels the area as a place where other groups perhaps discarded their unwanted cultural artifacts—or as a place of people who never invented anything significant or made any significant cultural or economic contributions to life in early Texas. However, nothing could be further from the truth.

Diverse groups of Karankawa lived along the Gulf Coast (map 2.1). They frequented the offshore islands on hunting expeditions and occasionally used them for permanent residence. Their activities extended along the coast in a narrow belt from the west side of Galveston Bay to near the present-day city of Corpus Christi.

According to early visitors to the area, plentiful food existed here, but residents faced serious seasonal limitations, especially if the food search was restricted to land. Local peoples, therefore, participated in seasonal hunting patterns to accommodate this shortage at certain times of the year. Little or no food was continuously available. Although summers along the Gulf Coast and on offshore islands were hot and humid, the winter months were often very dry, and supplies of drinkable water would become dangerously low (Guderjan and Canty 1989, 9). The Karankawas migrated from place to place every few weeks as food supplies dwindled. They returned to their favorite places year after year, much like other nomadic groups.

EARLY VISITORS AND SETTLERS

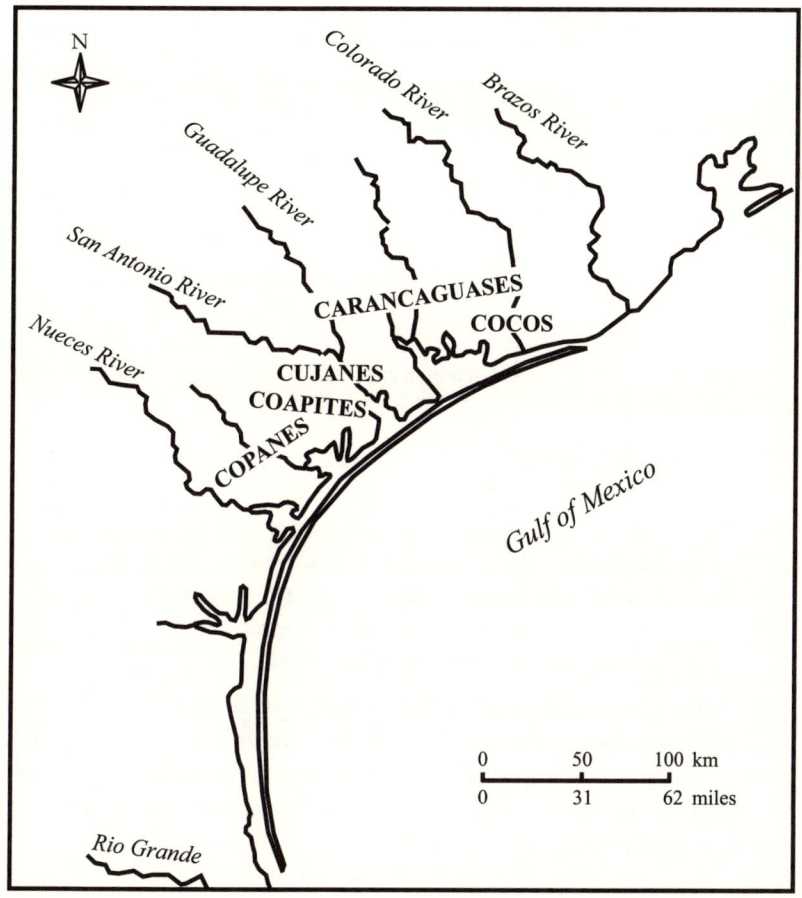

MAP 2.1. *Karankawa settlement on the Texas Gulf Coast. (Adapted from Ricklis 1996; cartography by Linda F. Prosperie)*

The Karankawas came into contact with the earliest Europeans on the Gulf Coast quite suddenly and unexpectedly. Their first encounter occurred in 1528 when the survivors of the Narvaez expedition washed up onto their beach. After a very turbulent time with Cabeza de Vaca and his compadres, more than a century and a half passed before the Karankawas came into contact with other Europeans. In 1685 the French explorer LaSalle established a small settlement, Fort St. Louis, on Matagorda Bay. The Spanish then decided to build missions among the Karankawa to safeguard their territory against the French. A few Indians responded and relocated to a mission settlement, but most resisted vehemently. By the end of Spanish rule in

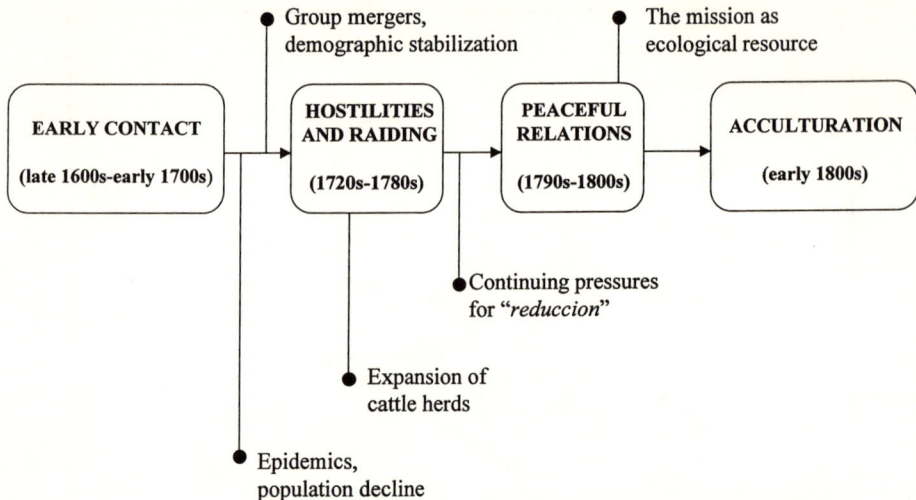

FIGURE 2.1. *Karankawa adaptive responses: impacts of European contact and civilization. (Adapted from* **The Karankawa Indians of Texas: An Ecological Study of Cultural Tradition and Change** *by Robert A. Ricklis. © 1996. By permission of the University of Texas Press)*

south Texas, the Karankawas had been nearly extinguished by warfare and by European diseases. Many died of smallpox, typhus, measles, and other diseases. A seminal study of the Karankawa people (Ricklis 1996) summarized the fatal impact of European contact and colonization on Gulf Coast aboriginal people (fig. 2.1).

EARLY EUROPEAN CONTACT

Galveston was nothing more than an elevated grassy sandbar 28 miles long and 1–3 miles wide when it was first seen by Europeans (Mason 1972, 22). Two Spanish explorers, Juan de Grijalva in 1519 and Alonso Alvarez de Pineda in 1530, were the first Europeans to see the island, as they passed by it. However, they didn't plan to visit for long (Miller 1983, 41). A few years later, in 1528, the first, equally reluctant group of "settlers" arrived on the island: eighty-plus individuals who were washed ashore in small boats adrift after the collapse of their failed settlement in Tampa Bay. The self-appointed leader of these Spanish adventurers, Cabeza de Vaca, and three of his companions—including the first African to settle in Texas, a man from Morocco named Esteban—survived the harsh winter of 1528 (map 2.2). Cabeza de Vaca not only survived but also carefully recorded a detailed historical account.

Not surprisingly, these four reluctant settlers considered their arrival on Galveston Island as a terrible misfortune. Among their many problems was the abundance of snakes on the island, which gave rise to the name Snake Island, or Isla de Culebras, bestowed by the earliest Spanish settlers. All the islands off the Gulf Coast of Texas were known collectively as Islas de Culebras because of the large number of rattlesnakes that lived under the driftwood and in the sand dunes. Cabeza de Vaca and his companions lived

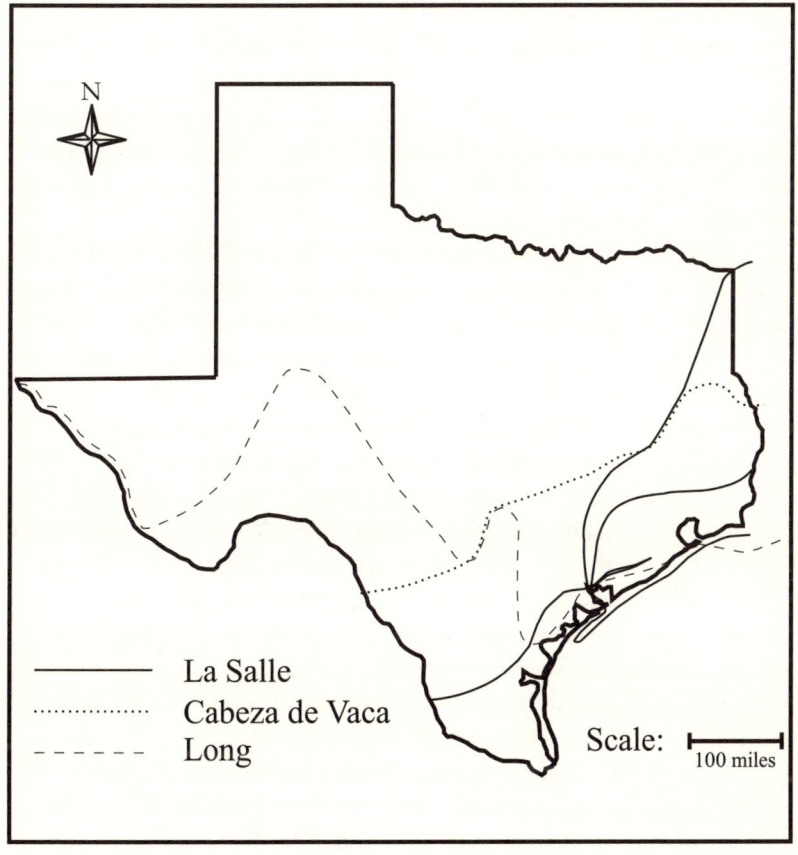

MAP 2.2. *Galveston's early European and North American visitors. (Adapted from Stephens and Holmes 1984; cartography by Linda F. Prosperie)*

among the Karankawas and other tribes on the mainland and the island for six years and then walked to Mexico with his three companions and Esteban or "Black Stephen" as he was called by many (Eisenhour 1983, 2).

More than a hundred fifty years after Cabeza de Vaca, the French were the next to arrive on the island. When La Salle discovered the island in 1685 as he searched for the mouth of the Mississippi River, he called it Ile Saint-Louis in honor of a famous king of France. It was known only by that name for about a century. According to the Sandusky map issued in 1845 (map 1.2), Galveston Island was originally two separate islands. The narrow strip of insular land was then separated by a bayou connecting the Gulf to Broadway and running though the island approximately where First Street is located today. The outer part of the island was named Culebra, while the island near the mainland was called San Luis.

The name Galveston was selected by the leader of the last Spanish exploration team, sent to the area in 1785, to honor Bernardo de Galvez, viceroy of Mexico. This name was officially drawn onto the first maps of the area when Spanish cartographers surveyed the Texas coast a few years later. Galvez, who was governor of Louisiana when Spain held it during the American Revolution, never visited the island.

The earliest account of French settlement comes from Commodore Louis-Michel Aury, who set up a base camp with his fleet of Gulf privateers in 1816. Aury, lured by the promise of riches to spend in New Orleans, arrived with about two hundred French war veterans and one hundred African adventurers recruited in Haiti (Faye 1931, 473). He also brought goods valued at more than half a million dollars, along with sixty thousand dollars in cash and a commission from President Bolívar to "cruise against the commerce of Spain" (*Morrison and Fourmy's General Directory . . . 1901–1902*, 11). Aury and his men quickly established a small settlement on the island and supported it with the help of a fleet of small sailing ships. These adventurers brought gold, silver, and other treasures stolen from Spanish ships, along with slaves, to the island, soon making it into one of the most prosperous and well-known places in the region. Slaves were transported to Louisiana, where they were sold for about one hundred forty dollars each. Aury sent some of the funds from this "export business" to Nacogdoches to support recruitment efforts for the revolution. Soon after the arrival of Aury and his pirate colleagues, another small fleet, loyal to the cause of freeing Mexico from Spanish rule, showed up in Galveston Bay. This fleet was under the leadership of Francisco Mina from Spain.

These were exciting times on Galveston Island, as revolutionaries from Texas, France, Spain, Mexico, Louisiana, Haiti, and beyond met and min-

EARLY VISITORS AND SETTLERS

gled to plan for the ultimate capture of Texas from Mexico—and to reclaim Mexico from its lingering Spanish dominance. Joining earlier settlers, Colonel Henry Perry and another hundred volunteers loyal to the Texas cause arrived in March 1817. In the spring a military coalition headed by Mina, Aury, and Perry sailed inland to help with the war effort. Their hastily organized attempt at collaboration, however, failed soon after their departure from the island, with Mina capturing Soto la Marina, Perry and his doomed men marching off toward Goliad, and Aury returning to Galveston. Even Aury left Galveston to establish another provisional government for France in east Florida (Faye 1931, 482).

In the meantime, the small party of pirates that remained at Aury's Galveston base camp had been taken over by the well-known brothers, Jean and Pierre Lafitte. The now notorious Lafittes and their adventurous colleagues from Louisiana, France, and Haiti had arrived on seven ships to take complete possession of the island. Despite Jean Lafitte's heroic patriotism during the War of 1812, the U.S. government forced him to leave his camp in Louisiana because of his covert operations in the Caribbean. When Galveston Island became available in 1817, Lafitte quickly made it his base of operations (Eisenhour 1983, 5).

The two French brothers supervised the construction of a fortified place of operation on two shell-covered ridges on the island near today's University of Texas Medical Branch. There they supervised the construction of a two-story headquarters, painted bright red and dubbed the Maison Rouge. A series of floorless, one-story huts centered on poles sunk into the sand were built nearby to house their associates. Along with these residences, a shipyard, slave market, and several food and drink establishments were also constructed, complete with billiard tables for additional entertainment.

The Lafitte economic system, first organized in New Orleans as early as 1804, was quite complex. The brothers purchased shares from other privateers and eventually acquired employees of their own, who were encouraged to own ships and even were allowed to use the Lafittes' base in exchange for some of their loot. In the four years when the Lafittes controlled Galveston Island, at least a thousand people lived there with them in a settlement officially owned by Spain, controlled entirely by the Haitian-born, and still quite French.

Early in 1818 a command under General Lallemand arrived in Galveston with a hundred more Frenchmen but no women. The intention was to establish a base of operation on Galveston Island to launch an attack that would capture Mexico for France. Some of these new arrivals ended up working for the Lafittes, while others traveled inland in an unsuccessful attempt to establish vineyards near today's town of Liberty, located on the Trinity River. Later, in 1821, a former U.S. Army surgeon, Dr. James Long, brought his family, including their child's African American nurse, and a small group of Mississippi volunteers to Bolivar, just across the bay from Galveston Island.

One story concerning Long's relationship with the few remaining Karankawas still living on the island illustrates the attitudes of American settlers and visitors. At that time, despite several centuries of domination and abuse by French, Spanish, Mexican, and American settlers and soldiers, a few Karankawas still resided on the west end of the island. These men had purportedly committed some "degradations" against Long's troop. This accusation led to Long and his men marching down the island to locate the Indians, who at the time were involved in a colorful dance ceremony near the place called Three Trees. The U.S. troops, coming upon the dancers by surprise, immediately attacked the ceremony and killed at least thirty Indians. One woman and her child were taken as prisoners. Seven Texas revolutionaries were wounded, but none was killed. Long, with his fifty-two men, sailed off to fight for Texas freedom in 1821, leaving his pregnant wife and young child behind. After Long was captured at Goliad, this early "freedom fighter" was taken to Mexico City, where he was shot.

One of the most interesting descriptions of life in early Galveston is contained in the diary of a colleague of Long, one Colonel Hall, who had the rare chance of interviewing Jean Lafitte:

> He found him to be one of the most prepossessing men he had ever met with, both in personal appearance and address. He was six feet and two inches high and his figure one of remarkable symmetry, with feet and hands so small, compared with his large stature as to attract attention. In his deportment he was remarkably bland, dignified, and social towards equals, though reserved and silent towards inferiors or those under his command. He received visitors with an easy air of welcome and profuse hospitality. He wore no uniform but dressed fashionably and was remarkably neat in his personal appearance. (*Morrison and Fourmy's General Directory . . . 1901–1902*, 12)

Despite Lafitte's promise not to attack American ships in the Gulf or Caribbean, a U.S. Navy warship sailed into Galveston Bay in 1821 and ordered the pirates to leave. This order launched an adventure tale fit for Hollywood. Nobody knows exactly what happened as the Lafittes and their employees prepared to leave the island, but many believe they hastily buried

EARLY VISITORS AND SETTLERS

their treasure near the original settlement on the harbor side of the island or the place "down the island" known as Lafitte's Grove or Three Trees. Leaving six men behind, Jean Lafitte set fire to his camp as he left, and all the buildings burned to the ground. Historical accounts of the ultimate destination of Jean Lafitte vary, but most agree that he ended up somewhere in the southern part of the Gulf and died on or near the Yucatán Peninsula.

Except for brief visits by supply ships carrying immigrants to Stephen F. Austin's new colony on the Brazos River in the early 1820s, Galveston Island was quiet for some years after this dramatic departure of the pirates. Austin finally visited Galveston Bay in 1825, calling it the best natural harbor in his new territory. Because of its strategic location for immigrant arrivals, Austin asked the Mexican government to allow him to add Galveston Island to his land grant. His request was denied. Soon afterward Mexico established a provisional port and customs house to help with the resettlement effort, thereby lending its support to Austin's efforts to populate the interior (Miller 1983, 57). A garrison of Mexican soldiers was sent to guard the customs house, forming the next wave of settlement. By 1832 there was a community of about three hundred people, mostly Mexican citizens, living on the island (Webb 1952, 662).

PERCEPTIONS OF EARLY VISITORS AND RESIDENTS

Galveston is a place where the perceptions of early immigrants often did not agree with reality. According to Jordan, "not infrequently, the settlers incorrectly perceived their environment, and made decisions based on these misconceptions. Indeed, one might well argue that the imagined character of the physical environment was of greater importance than the actuality for men act on the basis of what they think they know" (1980, 1).

Images and expectations of these newcomers who arrived at the Port of Galveston set the stage for tremendous efforts to control and change the physical environment of the Galveston Bay area in subsequent years. The alteration of natural surroundings, as well as the effort to force the landscape to conform to cultural ideals and to be amenable to human occupation, led to dramatic changes in the physical and environmental geography of this area of the Texas Gulf shoreline.

Europeans arriving in Galveston had conflicting opinions of its physical environment. Cabeza de Vaca had reported in 1536 that the Galveston area had no significant resources (Pearson 1994). Later, visitor Francis Sheridan's descriptions of Galveston also were quite disparaging:

The appearance of Galveston from the Harbour is singularly dreary. It is a low flat sandy Island about 30 miles in length and ranging in breadth from 1 to 2. There is hardly a shrub visible. . . . The nature of the soil on which Galveston is built . . . [is] a mixture of mud and sand generally up to the ankles of the pedestrian. This circumstance is productive of various complaints such as Yellow Fever, Ague, Rheumatisms &c, for when either from heavy fall of rain, or an irruption of the sea (a frequent occurrence during the prevalence of the "Northers") the ground gets thoroughly soaked the effect of the hot sun extracts such a horrible stench as no ordinary nose has smelt. (Pratt 1954, 31, 45)

N. Doran Maillard, writing in 1840, described Galveston as "an inhabitable sandbank" (Doughty 1987, 104). But many other early visitors had a more positive image of the island. Matilda Houstoun, for example, seemed thrilled to see an "immense flight of hummingbirds" that descended on Galveston Island in March 1844, while William Bollaert likened Galveston Bay in the 1840s to a "great nation of geese" (Doughty 1987, 39, 41).

Imported trees, such as the orange and myrtle, and nonnative flowers, such as roses and oleanders, contributed to positive visual images, reminding new arrivals of their former homelands. Matthew Maury described the Gulf of Mexico as a Mediterranean environment with a temperate climate characterized by perpetual summer (as described in the *Galveston Daily News*, 11 May 1894, 1). Immigrants from Southern Europe, in particular Greeks and Italians, found Galveston quite similar to their native lands. In 1889 Maury described a seemingly Mediterranean scene, a fleet of a hundred small boats in Galveston Harbor where men sold their daily catch of fish, oysters, and other goods (Maury 1849, 516).

The climate of Galveston Island, then, had both supporters and detractors. As Jordan (1980, 9–13) points out, in general, the Europeans correctly interpreted temperature but evaluated the climate as consistently more tropical than it was in reality. Maillard disparaged the Texas climate as being subject to sudden changes—hot summers followed by storms and severely cold winters—and he blamed rapid climatic changes for frequent outbreaks of fever. Sheridan described a norther that "blew like blazes" and was "shockingly dirty" (Pratt 1954, 9). The frequent hurricanes that made landfall in Galveston were consistently ignored until the storm of 1900. In 1818 Galveston was oversurged by a hurricane. Wrecks from this storm were found five miles inland in the 1820s (Jordan 1980). Hurricanes also devastated the area in 1837 and 1842.

Early accounts of the environment of Galveston ranged from extremely

EARLY VISITORS AND SETTLERS

positive to extremely negative. With due allowance for exaggeration, both viewpoints contain kernels of truth. Because Galveston is a barrier island, a superficial examination can easily give rise to a belief that it is a desolate place. Unfamiliarity with native species and ecosystems reinforces this viewpoint. At the same time, human affinity for a place and the depth of the human desire to create a home can lead to a positive image, as held by many who lived in and visited Galveston. One mid-nineteenth-century visitor, Wilhelm Steinert, expressed the usual "mixed bag" of both extremely positive and negative images associated with such a foreign environment:

> It is laid out in a regular grid on the eastern end of the very narrow, 15-mile-long island of the same name, and has broad main streets. The houses, although made of wood, look quite comfortable. The beautiful broadleaf trees along the streets as well as in the gardens, give the city a very pleasant appearance. The oleanders especially, with their countless blossoms, catch the immigrant's eye. I saw figs, pomegranates, lemons, sour oranges, oranges, and the like in some gardens. . . . In general, the surroundings of the city offer a dismal sight. Grass grows in shabby abundance on heaps of shell dust. Several swamps spread evil odors. One sees no wooded sections and reputedly only three trees grew on the island originally. Stovewood drifts in from the Gulf. (Steinert 1999, 39)

SURVIVING LIFE ON THE EDGE OF TEXAS

Despite often positive perceptions of opportunities offered by Galveston Island's rich natural resources, life there was not always easy. Finding enough edible food for survival posed few problems for the earliest inhabitants of the island. As mentioned earlier, Native American hunting parties found many kinds of animals and plants to satisfy their hunger and nutritional needs among the island's sandbars and offshore in the Gulf waters. The Karankawas even found ways to cope with the mosquitoes, fleas, ticks, and the other challenges of a tropical climate; for example, they covered themselves with tar to keep the bugs away.

Later arrivals from Europe and North America did not fare as well. This was not due to the lack of local food, for food was brought in on ships and sold at daily meat and vegetable markets. According to Wheeler, both beef and fish were plentiful and cheap, although pork and mutton were both very expensive. Farmers had a difficult time raising vegetables in the island's sandy soil. Drinking water on the island also posed a problem, because of both its scarcity and its foul taste. Most settlers relied on rainwater trapped in barrel cisterns for their personal supply. In addition, fresh water was available from streams flowing into the Gulf just across the bay, although it was difficult to transport this water to the island.

The biggest problem for most newcomers was coping with the insects and the heat. According to G. M. Addison, a resident of Galveston in the 1840s, "you sometimes think you are troubled with fleas, but you have no cause for complaint, and were you to spend a few nights in Galveston, you would be presented with the fact in a most demonstrable light.... The mosquitoes however are more formidable, and the poor fellow who has to sleep without a 'bar' is indeed in a 'bad fix' and really is entitled to sympathy" (Addison 1845, 1).

Insects were not the only challenge of life on Galveston Island. While its excellent natural harbor and strategic location may have attracted explorers and settlers, its hot, humid island environment, along with poor sanitation and isolation from medical facilities in its early years, made disease a constant problem.

Early written accounts detailing the unhealthfulness of the island are found in numerous letters home and in settlers' and visitors' diaries. Even Austin, who encouraged early Texan settlement of Galveston Island as the perfect natural harbor for his new colony, noted that the "situations back from . . . the coast, are remarkably more healthy" (Barker 1929, 143). In Sheridan's diary of his 1839–40 visit, he wrote about Galveston's perceived unhealthfulness: "Galveston must be very unhealthy, and require the greatest attention and caution . . . to prevent a yearly visit of the yellow fever as in New Orleans. . . . The lower country, from the Trinity to the Colorado, is as sickly as the most unhealthful portions of Louisiana. . . . The country becomes healthier at any point as you recede from [the] Gulf" (Muir 1958, 120, 132).

Strangely enough, a few residents said exactly the opposite about Galveston's health situation. One early visitor reported that "the island has the reputation of being healthy on account of the constant and refreshing breezes from the ocean, and my own experience during the sickly season fully confirms the prevailing opinion. During the summer, many invalids collected at this point from every part of Texas to embark for the United States and in a short time all revived under the salubrious influence of the climate" (Muir 1958, 5).

Even several of the island's earliest physicians praised its healthfulness. One of these was respected pioneer doctor Ashbel Smith, who advised his patient, Republic of Texas president Mirabeau B. Lamar, to recuperate in

Galveston following a serious bout with "bilious fever" (Silverthorne 1982, 58). Later that fall, following one of several serious epidemics of yellow fever in Galveston, Dr. Smith, in a spirit of enthusiastic boosterism, told a friend, "If you have any invalids send them among us. An abundance of seafood—an atmosphere unrivaled for balminess and salubrity—and novelty of scenes and the excitement of virgin country are at the command of the dyspeptic. For rheumatics and consumptives, the climate is particularly congenial" (Smith Papers, 21 October 1839).

No disease brought more fear and more deaths to Galveston's early residents than yellow fever. At least seven major epidemics of this killer disease swept through the island's population before 1860, killing more than twenty-three hundred people. Yellow fever was a common disease in the entire Caribbean region after the mid-seventeenth century. The yellow fever virus was first brought to the area on African slave ships and then was easily transmitted by mosquitoes, which flourished in the South. By the eighteenth century yellow fever had arrived in North America, establishing itself most frequently and most persistently in Gulf Coast ports.

The first recorded yellow fever epidemic struck Galveston in 1839, claiming 250 lives (Miller 1983, 69). Hayes described this event: "In the spring and summer of 1839, Galveston presented a scene of active progression. Houses were being erected as if by magic. But this busy scene of progressive life and animation was suddenly paralyzed and the energies of the people were instantly numbed by a dreadful fear, and friend looked into the face of friend, neighbor into the face of neighbor, with the fearful inquiry of 'Who next?' An epidemic had fallen upon them and was decimating their ranks with a fatality more dreadful and irresistible than war" (1879 [1974], 272).

This epidemic gave citizens an early taste of how fragile life on a sandbar could be for its human inhabitants. And yet despite the many challenges faced by early residents on Galveston Island, they continued to come—and they came by the thousands—making Galveston the largest city in Texas at the time of its first census. According to customs house records kept at Rosenberg Library in Galveston, the following global connections were a regular part of Galveston's import-export business as early as 1835: New York, New Orleans, Mobile, Baltimore, Vermillion Bay, Savannah, Philadelphia, and Wilmington. In 1839 ships arrived at Galveston Harbor from places such as Asabove Plus, Pensacola, Havana, Richmond, Bath, Bangor, Portland, Kingston, Georgetown, Boston, and Nassau.

Birth of a Global City

The vessel *Galliott Flora* from Liverpool docked at the crowded Galveston Wharf on May 2, 1849. At first glance, the ship appeared overloaded with its busy crew and 165 German passengers. These soon-to-be Texans were reported to be members of the "wealthy class," in a front-page article in the *Galveston Weekly News* the following week. Ship's manifests listed them as farmers, lawyers, gardeners, ladies, saddlers, butchers, carpenters, merchants, servants, bakers, hunters, masons, blacksmiths, and shoemakers. Although some families stood together as the ship threw anchor, an onlooker standing on the dock watching this international scene would have noticed that most of the ship's passengers were young men. The youngest passenger to disembark (no doubt in the arms of his mother) was six-month-old Anton Mochichek; the oldest was sixty-year-old Eva Gimpel.

Young Joseph Ellenburger and his thirty-three-year-old wife and four-year-old daughter were typical passengers on the *Flora* as it docked in Galveston that warm spring morning. The Ellenburgers had decided to leave their farm in the Lippe-Detmold state of Germany to travel to Texas to begin a new life. The route taken by this young family fit the typical Germany-to-Texas-by-way-of-Liverpool model perfectly. Their home had been near well-established immigration routes to French, Dutch, and English ports.

Even the Ellenburgers' first few days in Galveston would somehow feel a lot like home. Although they spoke little English, the Ellenburgers clearly understood the language spoken on the street, inside the stores where they bought their food, and in the saloons where Mr. Ellenburger drank his beer. Indeed, they must have been quite surprised to discover that much of the time they were surrounded by other newcomers who spoke the same dialect as friends and family they had so sadly left behind.

Despite these cultural connections, the hot, humid weather in Galveston certainly didn't feel like home. Yet somehow the Ellenburgers felt strangely secure. In these first heady days, this young immigrant family probably never imagined that they would soon leave this German-dominated city and move on before the first city directory was published less than ten years later. No doubt most of the German passengers who left the *Flora* on that hot, early summer morning were also completely unaware that political change was on the horizon. This change would seriously affect their lives and the lives of native-born Americans in Galveston alike.

Less than a week after the *Flora* arrived at the busy Galveston docks, the editor of the *Galveston Weekly News* published his regular diatribe against the U.S. Constitution, calling it a document "of compromise . . . that should be compromised or changed from time to time." Other front-page stories in the May 14, 1849, issue of the paper expressed outrage at northerners who refused to return runaway slaves to their owners. These attitudes concerned major political and social issues that would come between many of the new immigrants and their southern neighbors — and would help shape decisions that would affect the Ellenburgers some years down the road. Despite the uncertain political climate and the oppressively humid air, however, after more than three crowded weeks aboard the *Flora*, Joseph and his wife and young daughter were more than eager to begin their new and still quite uncertain life as Texans.

No documented letter or journal tells precisely how the young German immigrant Joseph Ellenburger and his family reacted to the heat, humidity, or political climate on the crowded Galveston docks in May 1849. What little we know comes from examining ship passenger lists and city census records. But we can be sure that many other immigrants from Germany and elsewhere experienced some of these same confusing feelings of both hope and despair when they stepped off the ship onto sandy Texas soil for the first time.

THE TEXAS REVOLUTION AND AFTERWARD

Almost fifteen years before the *Flora* sailed into the harbor at Galveston in 1849, Stephen F. Austin had become aware of the excellent natural harbor on the bay side of Galveston Island, adjacent to the southeast corner of his land grant. In 1825 Austin petitioned the Mexican government to allow him to establish a port on the island, but his request was refused. The island then remained uninhabited by Europeans or Americans until a Mexican customs house was established there in 1830, which attracted a small settlement. According to the travel notes of naturalist John James Audubon, five years later the settlement still had only a miserable collection of huts, prisoners, and chickens, plus one pig and one dog.

Mexican residents abandoned the small settlement during the war for Texas independence from Mexico. At the end of the war Galveston Island served briefly as the capital of the republic of Texas, and from 1835 to 1837 the Texas Navy was based on Galveston Island at Fort Point near today's Coast Guard station (Miller 1983, 58). The fledgling Texas provisional political leaders also took refuge on Galveston Island in 1838 in an effort to escape capture by Santa Ana's forces. The first president of the Texas republic, David Burnet, and his staff ultimately heard about the Texan victory at San Jacinto from messengers who rowed across the bay to report the news (McComb 1983, 10).

Microfilmed records from the Texas-Mexico War at Galveston's Rosenberg Library list 1,844 prisoners of the Mexican government who were bought to the island. The records also give the names of all Texans killed by Mexicans up to 1842; the names of those ordered shot by Santa Ana; escapees; those liberated by Santa Ana; and prisoners who were marched to Mexico. It is beyond the scope of this chapter to provide details of these gruesome lists of the living and the dead, but these data lend credence to the importance of Galveston as an early port city with well-established linkages to the outside world. A visitor to the island in 1837 recorded this description of the prisoners who remained in Galveston: "They were cantoned in small round huts, formed of square pieces of turf, piled to the height of six or seven feet.... [T]he naked condition of many of the prisoners showed that their sufferings, which must be considerable in a climate where the change of temperature is sudden and the cold oftentimes severe, created but little sympathy in the bosoms of its conquerors" (Muir 1958, 7).

According to customs house records, in 1835 ships arrived in the harbor of Galveston from New York, New Orleans, Mobile, Baltimore, Vermilion Bay, Savannah, and Philadelphia. In 1839 ships came to Galveston from those places as well as from Pensacola, Havana, Richmond, Bath, Bangor, Portland, Kingston, Georgetown, Boston, and Nassau.

Records kept by Dr. Ashbel Smith, respected physician of the republic of Texas, in 1839 revealed that Galveston had grown significantly from the single house he had observed only two years earlier. Galveston had turned into "a bustling town, 'twice as large as Salisbury (which had about 1,000 inhabitants), with twenty-five or thirty sailing ships in the harbor at all times and a dozen or so steamboats moving in and out during the day.' The change, he wrote Charles Fisher, was like 'the magic of the Arabian Nights stories'" (in Silverthorne 1982, 58).

Although the majority of Galveston Island's earliest visitors arrived in the city from the port of New Orleans or by water from ports elsewhere, some did travel to the new city by land. According to the author of *Prairiedom*, writing in 1848, "Galveston is the most important seaport in Texas." This account described the travel route from Gaine's ferry, entering Texas by way of Shreveport across the "bloody prairie," the Sabine River, and finally connecting with the Camino Real at Nacogdoches. After passing through the Coushatta Indian village in east Texas, this traveler then crossed the Trinity River:

> We crossed the Trinity (Trinidad) at Robin's Ferry . . . the Trinity affords the best steamboat navigation in Texas. Boats have already ascended as high as Elk Heart, and in good stages of water can easily go to Two Forks, so called in the mineral regions some two or three hundred miles above. We made the trip up the river, some four or five hundred miles, and found few impediments to navigation . . . about the middle of April, we took passage on a steamer for Galveston. We descended Buffalo Bayou, a deep but narrow channel of easy navigation. We reached Galveston the next day at noon" (White 1965, 17).

Clearly, many difficulties faced travelers to Galveston who came by way of land and rivers. By far the most common route to Galveston was by ocean steamer from the German port of Bremen by way of New Orleans. A German named Wilhelm Steinert coincidentally arrived in Galveston by taking this route on exactly the same day as Joseph Ellenburger. Steinert saw the *Flora* also arriving at Galveston's harbor; the *Flora* had left Bremen on exactly the same day as Steinert's ship fourteen days earlier.

Steinert's description of early Galveston is instructive: "Only 4,000 to 5,000 people living here, one-third are German. For this reason one can get along very well with the German language, although I did not find true . . . that people almost hold it against you if your speak any language other than German. The American language and customs play the chief role. . . . There are several churches here, especially the Catholic one, built in good style. Several schools have been established. The German school provides free instruction for children of parents without means . . . instruction is given in German and English" (Steinert 1999, 39).

One of the settlers who most influenced the layout and design of the early town plan of Galveston was a French Canadian named Michel Menard. This foreign visitor shared Austin's vision of the potential for developing a large port on the island. But when Menard arrived in Galveston, eager to buy land and develop the island into a city, he was told that Mexican law no longer permitted non-Mexican foreigners to own land in Texas. Still anxious to buy land on the east end of the island, Menard asked his friend, Juan Seguin, who was a native of Mexico (and therefore eligible to receive free land), to ask for the island as compensation for having fought on the side of Texas against the Indians.

After much discussion, Seguin was granted a "league and a labor" at the eastern end of Galveston Island.* Menard surveyed the land and bought it in partnership with Thomas F. McKinney, a business associate of Austin's agent, Samuel May Williams. This group of investors and land developers organized the Galveston City Company in 1836 (Frantz 1988, 92–93).

By the time this transaction took place, the republic of Texas had become a reality, and Galveston was its only port. However, this political change resulted in an investigation into the legality of Sequin's land grant. The Texas Congress insisted that the land be sold to the highest bidder. In the end, Menard and McKinney's bid of $50,000 was approved. The funds were offered to the Texas Congress in the form of clothing and provisions for the army, thereby confirming Menard's claim and, at the same time, officially creating the Galveston City Company (Wheeler 1968, 70). The republic of Texas established its own customs office inside the old Mexican customs building at Galveston and built a fort on the east end of the island (Miller 1983, 59).

Except for shifting sands and areas with poor water drainage, flat, unimpeded Galveston Island was relatively easy to survey. The City Company platted the town according to a plan developed by New York surveyor John Groesbeck in 1838. It was laid out in a standard grid with blocks divided into fourteen long, narrow lots. Streets running east-west were named alphabetically, while those running north-south were numbered. The incorporated city was located from Seventh Street west to Thirty-first Street and from the Gulf northward to the channel of the bay. The Galveston City Company donated four blocks (nos. 100, 101, 160, 161) to the city for burial of

---

*After Mexico won its independence from Spain in 1821, new laws were put into effect that controlled immigration and land settlement in places such as Galveston, now owned by Mexico. These new laws invited Catholics to settle in Mexico, provided for the employment of agents called *empresarios*, introduced families in units of 200, and defined the land measurement in terms of labors (177 acres each) or leagues (4,439 acres each). Those who agreed to farm the land were promised at one least one labor of land, those who raised cattle were given a league, those who were both farmers and ranchers were given a labor and a league. Settlers were free of titles and taxes for six years and paid only half payment for another six years (Webb 1952, 2).

the dead. Before this dedication, bodies had been buried in sand hills in the "back of the city"—but the shifting sand dunes kept exposing the bodies. Francis Sheridan, an early visitor to the island, described the early cemeteries as being "a little way out of town among some sandy hillocks through which runs a public road. It is unenclosed on either side and entirely deficient in 'storied urns or animated busts' as may be supposed the prevailing fashion being small pieces of Board whereupon the initials of the departed are carved . . . it merges into a swamp and some of the graves in consequence filled with and destroyed by water" (Pratt 1954, 48–49).

Various measurement systems were used by surveyors in different parts of the city, depending on local topography and hydrology. For example, between Seventh and Twenty-ninth Streets, each block was divided into fourteen lots, seven on the north and seven on the south side of the block, with the standard 20-foot-wide alley running east and west. West of Thirty-third Street, blocks did not have alleys. Beyond W Street, west of the official City League, all land was surveyed into 10-acre lots in 1837. Few lots on the west end were sold initially, even though the price was only two or three dollars an acre. One can only imagine the surprise of these early surveyors if they had known that prices for a small lot on the island's west end would cost over $100,000 in the year 2000.

Two developers sold the first lots in April 1838. A few lucky investors were given water frontage on the channel side of the island in exchange for construction of wharves and warehouses. The City Company sold 700 lots the first year for an average price of $400 per lot. One of the first buyers was Joseph Osterman from the Netherlands; he bought a corner lot at Market and Tremont Streets and built a store there (Miller 1983, 61). Within a few years, by December 1837, Galveston had seven semipermanent houses and regular activity at the port. Charles Morgan had established regular steamboat service between Galveston and New Orleans, launching the first ship of the Morgan Line. Galveston County was organized in 1838 with Galveston as the county seat. That same year, the city's first newspaper, the *Civilian and Galveston Gazette*, was published, followed two years later by the better known and much longer lasting *Galveston News*.

The city was incorporated in 1839 and divided into three wards. Its first mayor, John M. Allen, a survivor of the Texas Revolution and a native Kentuckian, was no doubt quite proud of the city's thousand-plus population, its well-organized street plan with over two hundred fifty houses, and its busy port and well-stocked wharves. During the summer of 1839 Galveston started its first serious period of growth. Gail Borden, collector of customs, proclaimed proudly that Galveston would surely emerge soon as the "New

TABLE 3.1. Ships from Bremen and Other Western Ports Arriving at Selected U.S. Ports, 1841–47

Port	1841	1842	1843	1844	1845	1846	1847
New York	38	41	39	56	75	79	107
Baltimore	36	35	37	36	52	50	48
New Orleans	16	28	29	36	56	45	35
Galveston	0	3	8	7	19	33	4
Total	90	107	113	135	202	207	194

Source: Adapted from Struve 1996, 4; from Pitsch, *Die wirtschaftlichen Beziehungen Bremens zu den Vereinigten Staaten*, 219.

York of the Gulf" (Wheeler 1968, 70). Despite the terrible hurricane in the fall of that same year and a yellow fever epidemic the following year that eliminated more than 10 percent of the city's population, Galveston not only survived but continued to grow. Tens of thousands of newly arriving immigrants from Europe added to the city's population, creating a cosmopolitan city with links to both the interior and the outside world. These newcomers arrived in sailing vessels and steamships, the number increasing rapidly after the mid-1840s. Many, like Steinert, came from the German port city of Bremen. Table 3.1 compares Galveston to other port cities in the United States as a port of entry for ships with immigrant passengers between the years 1841 and 1847.

The experience of arriving at the port of Galveston was described by Dr. Ferdinand Roemer, who arrived at the new city from New Orleans in 1847: "Shortly after this we saw a white streak, which indicated breakers at the bay laying at the entrance to the harbor. We steered a straight course toward it, and soon found ourselves in the midst of the turbulent waves. A few moments later we had successfully surmounted this obstacle, and sailed around the northern point of the island which was separated from another strip of land by a narrow channel. Thereupon, we landed immediately at the City of Galveston" (Mueller 1935, 39).

The first German newspaper in Texas, the *Galveston Zeitung*, was published between 1841 and 1855 in Galveston. In addition, the city claims these other firsts in the state: chamber of commerce, private bank, telegraph, jewelry store, national bank, electric lights, training school for nurses, brewery, golf course, and medical college. By the year of the first federal census in

1850, Galveston had become a vibrant city with more than three thousand people and a bustling seaport (Frantz 1988, 89, 93).

Galveston had been an international city from the very beginning. In the years following exploration by Spanish adventurers and settlement by French pirates, the city continued to turn its face toward the sea—a trend that continued throughout the nineteenth century. As early as 1857 Galveston's first city directory listed foreign consulates in Galveston representing eleven countries: England, France, Mexico, Prussia, Hamburg, Austria, Saxony, Bremen, Netherlands, Nassau, and Spain. A hand count of all foreign-born residents listed in the 1850 census provides ample evidence that the city's residents were also increasingly global.

EARLY EUROPEAN SETTLEMENT IN GALVESTON

By the late 1840s the city of Galveston had evolved into a place where people from diverse cultures and ancestries lived side by side. An excerpt from a letter written to potential German immigrants by an anonymous visitor to Galveston in 1837 had this to say about its possibilities for settlement: "There is nothing Texas deserves and actually stands in need of as much as *population* . . . it cannot be a matter of wonder that the most alluring prospects are held out to the many found in every country who are always willing to change their situation to improve their condition in life" (Muir 1958, 163).

Predeparture information was directed in particular to large groups of immigrants leaving for Galveston from the German port of Bremen. Although the most common image of Germans in Texas is rural rather than urban (see Jordan 1966), Germans were by far the largest group of early Europeans to settle in Galveston. A count of all foreign-born names listed in the 1850 manuscript census determined that Germans were the largest group of immigrants living in the city that year. Many more arrived in the decades after the Civil War. Thousands of others, on their way to inland settlements such as New Braunfels in the Texas Hill Country, stopped briefly in Galveston to prepare for their journey inland.

German immigrants were drawn to Texas by the promise of cheap land and economic opportunities. Their desire for religious and political freedom also encouraged many Germans and other Europeans to leave their homeland, especially during and after the revolutions of 1848. "In particular, the persecutions to which they were subjected in Prague and Budapest following the revolution caused many Jews to migrate to Galveston because it was reported that the Texas island city offered them political, economic, and religious freedom" (Kisch 1949, 185).

A list of advantages and disadvantages for Europeans who emigrated from their homelands to Galveston and beyond was compiled by one of the island's earliest German visitors in 1847. The advantages included:

- Mild and excellent climate
- Excellent prairie soils; thick, tall grasses; and ample level land for successful cattle raising
- Natural conditions for successful agriculture—no woods to clear, fertile soils, ample water, no fertilizer necessary
- Cheap price of land
- The sparse population, many of them immigrants, providing a social and cultural context for living among one's own countrymen. (Mueller 1935, 26–32)

According to this same writer, disadvantages of the new land included:

- The shock of living in a climate so completely different from that of Germany, which often caused new immigrants to experience sickness and weakness
- Diseases like bilious and malaria fever, which often struck new residents who lacked natural resistance to them
- Dysentery, ague, and malaria, all distributed widely along the Gulf Coast
- Living with the imperfect communication systems between Galveston and the inland of Texas—the lack of roads and highways, railroads, and few natural water courses for navigation
- Galveston's location in a slave state, which meant that new arrivals, lacking slaves, were often viewed as inferior in terms of social class and economic potential. (The writer also noted that the labor of the free man was not nearly as respected by native-born residents as the work of a slave.)

This German traveler clearly had mixed feelings about German settlement in Galveston: "When weighing the advantages and disadvantages of settling in Texas, especially as compared to settling in the more northern states, no doubt the advantages in Texas far outweigh the others when not taking the danger of health into account. If one, however, places a greater value upon health and there prefers to live in a more healthful climate where

BIRTH OF A GLOBAL CITY

he can also establish himself, but with greater efforts, he had better select the Western or Northern states, particularly Wisconsin, Iowa, Illinois, and Missouri" (Mueller 1935, 29-30).

Despite these well-publicized disadvantages of Germans relocating to Texas, German migration became important as early as the 1830s when Friedrich Dirks, known as "Ernst," first settled in a rural area roughly halfway between Houston and Austin during Mexican rule (Jordan 1966, 89). After Ernst sent a letter home to friends expressing the favorable qualities of his new home in Texas, his ideas were eventually reprinted in newspapers and in immigration brochures. Early emigrants who left Germany to resettle in Texas drew the attention of the Adelsverein, a German organization interested in developing a German colony abroad. "Between 1844 and 1847, thousands of Saxon and Hessian peasants under their direction were settled in the Hill Country west of Austin creating a second, western German focus in the state" (Jordan 1989, 89). The first immigrants to reach the port of Galveston under the protection of the Adelsverein arrived in Galveston in July 1844, and the last one passed through the city in 1846. A third node of German settlement in early Texas was established by Henri Castro and a group of German Alsatians at today's Castroville just west of San Antonio. Terry Jordan has mapped out the areas of origin of German settlers in Texas (map 3.1).

A hand count of the list of foreign-born residents' names in the manuscript census of 1850 reveals that more than 40 percent were German or Prussian. Adding to the German population in the city of Galveston were other German families and individuals who were passing through on their way inland. To raise enough money to make the long trip to the hill country and other parts of central Texas, some stayed on and used their skills in construction, iron work, masonry, and mechanics to help build the new city. According to ships' records, others arrived as highly skilled tailors, millers, and shoemakers. As such, their "ideas, tastes, and talents were brought to bear upon architecture, street and park construction, music appreciation, and to some extent, upon the intellectual life of the city" (Fornell 1955, 15). Of particular interest were changes in Galveston's original southern food preferences. Most noticeable was the replacement of southern cornbread with German-preferred wheat bread on the island.

Newspaper editor Ferdinand Flake represents one piece of the German mosaic in Galveston. Deciding to leave his homeland in Germany when he was only eighteen years old, Flake earned his living selling cigars, owning his own mercantile business, and eventually, in 1857, launching the first German-language newspaper in Texas. Flake's paper quickly achieved a

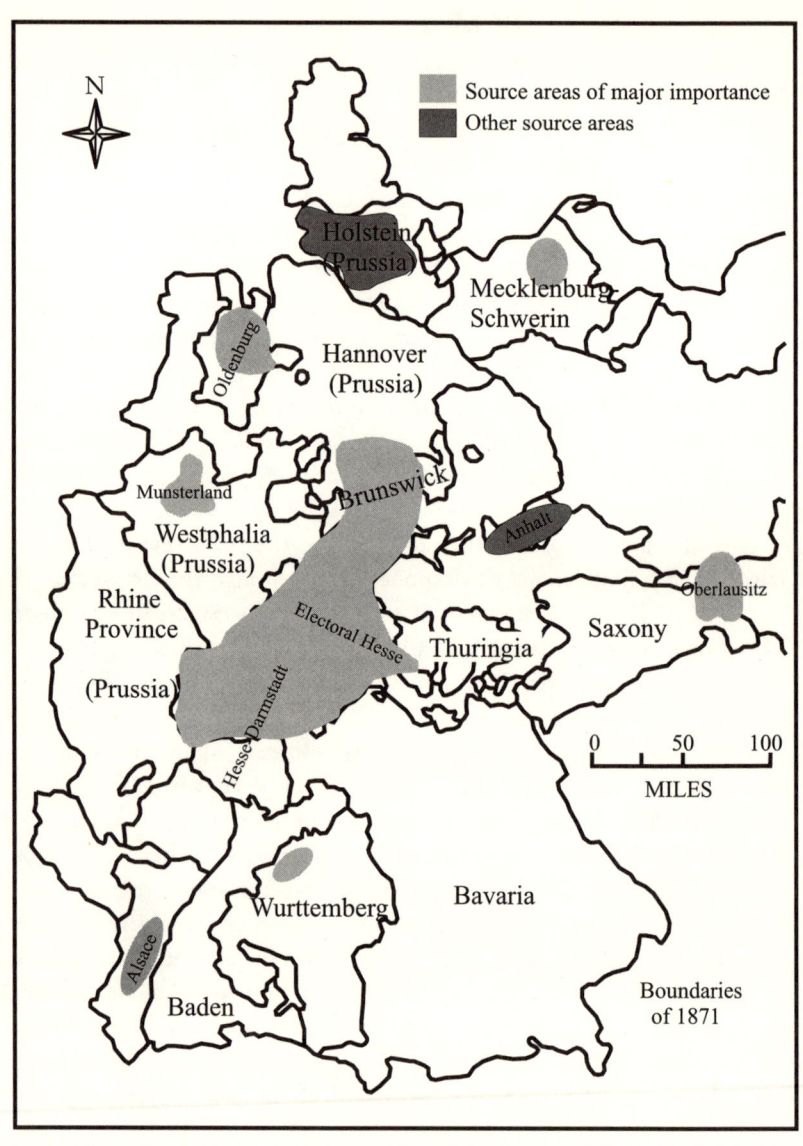

MAP 3.1. *Areas of origin of German settlers in Texas immigrant residential pattern, 1850. (Adapted from Jordan 1966; cartography by Linda F. Prosperie)*

larger circulation than either of the other two newspapers in Galveston, the *Galveston Daily News* and the *Civilian*. Flake was more than an editor, he was a strong pro-Unionist, humanitarian, and social reformer. He and his brother Adolph organized the effort to feed families on the island during the federal occupation in the Civil War.

Germans who remained in Galveston emerged among the most industrious and stable people in the community, as individuals rapidly rose to the social status of major merchants and civic leaders (Wheeler 1968, 78). This large immigrant group supplied the residents of Galveston with most of their fresh vegetables, fruits, and dairy products from market gardens and dairies outside the city limits (see Jordan 1966, 75, 91). In addition, they owned many of the city's businesses and had a lasting impact on its cultural landscape (map 3.2). Occupations of the German residents in Galveston in 1850 are summarized in table 3.2.

German "landscape signatures" in the city are still evident even to the most uninformed tourist. For example, on a sunny springtime afternoon in May 1998, the public was invited to celebrate the rededication of Garten Verein, a historic German meeting hall and still one of the most beautiful reminders of the German presence on the Texas Gulf Coast. This event marked the end of a million-dollar campaign to restore this octagonal structure as a social and cultural gathering place with a surrounding park for public use (fig. 3.1). The pavilion was open all summer in the old days and provided music, bowling, food, drink, and sometimes, even romance.

Despite their relative stability as successful colonizers, numerous tales have been told about the Germans who settled in Texas in the mid-nineteenth century. Stories about their antislavery attitudes, their wealth and power, and their stoicism abound. For a few of the Germans in Galveston, some or perhaps all of these stories were true. The majority of Germans arriving in Galveston Harbor were either skilled craftsmen who were accustomed to urban living or rural peasant farmers; at least a few of these early pioneers were unquestionably wealthy (Fornell 1955, 15). Other stories about Germans in Texas suggested that (1) Germans did not own slaves; (2) they were in favor of the abolitionist cause; (3) they were morally opposed to slavery; and (4) they were staunch pro-Unionists during the Civil War (Jordan 1989, 92).

In a study of the impacts of this immigrant group in Texas, Jordan provides data that thoroughly debunk each of these myths (1989, 92–96). First, he explains the origin of the German abolitionist stereotype. While a few Germans were antislavery—and many of these more liberal and urbane set-

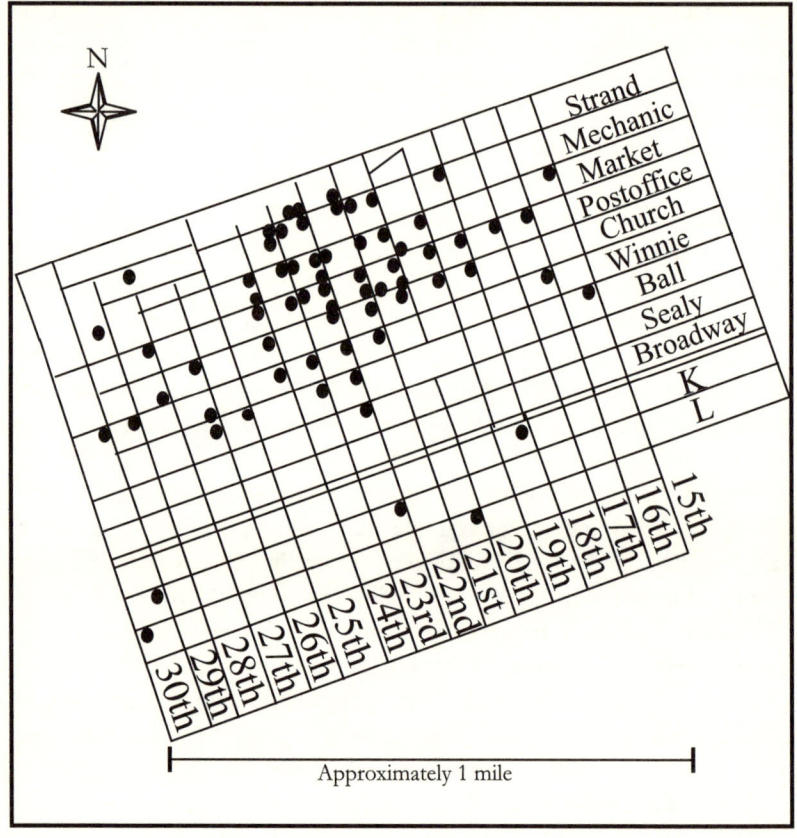

MAP 3.2. *German businesses in Galveston, 1857. (Cartography by Linda F. Prosperie)*

tlers lived in Galveston—they were a small elite minority of educated liberals such as Ferdinand Flake, who published his antislavery, pro-Union views in Galveston and in other Texas newspapers. However, according to Jordan, there is little evidence that the majority of Germans who relocated to Texas agreed with the antislavery position of the German intellectuals. Many of them reacted to the "abolitionist declaration of the 1854 convention by issuing a statement . . . recommending to [their] German countrymen to countenance and suppress all attempts to disturb the institution of slavery" (1989, 92).

Other Germans supported the secessionist cause before and during the Civil War. In Comal County, where 80 percent of the population was German, three companies of young German men were sent to fight in the Civil

War on the side of the South, and 74 percent of the electorate voted in favor of secession (Jordan 1989, 95).

Stereotypes to the contrary, some Germans in east Texas, where cotton growing required large inputs of labor, did own slaves. Jordan found that at the time of the first Texas census in 1850, 9 percent of all German farmers in the eastern part of the state owned slaves. "Clearly, Germans were acquiring slaves, particularly in the 1850s, and had abolition not come following the Civil War, the eastern settlements would almost certainly have become firmly enmeshed in the coastal southern slave-cotton system" (Jordan 1989, 94).

Other Europeans joined the Germans in the settlement of Galveston in the years before the Civil War. The manuscript census of 1850 lists, in decreasing order, residents born in Ireland, England, France, Scotland, Switzerland, Italy, Canada, Mexico, Spain, Denmark, Sweden, Poland, and Wales. Residential patterns of these foreign-born European settlers, based

TABLE 3.2. **Occupations of the German Residents of Galveston, 1850**

laborer	71	mechanic	4	barber	1
carpenter	43	physician	4	bookbinder	1
drayman	22	seaman	4	bracer	1
shoemaker	19	cooper	3	brewer	1
clerk	15	gardener	3	cabin boy	1
merchant	14	milliner	3	engineer	1
cabinetmaker	13	minister/preacher	3	fisherman	1
tailor	12	musician	3	hunter	1
apprentice	9	saddler	3	iron molder	1
butcher	9	boatman	2	magistrate	1
baker	8	confectioner	2	music teacher	1
printer	8	goldsmith	2	newspaper editor	1
apothecary/druggist	7	grocer	2	oysterman	1
barkeeper	7	milk seller	2	piano maker	1
boarding house keeper	5	sailmaker	2	servant	1
cigar maker	5	ship's captain	2	ship's carpenter	1
wheelwright	5	tanner	2	ship's cook	1
blacksmith	4	teacher	2	ship's mate	1
bricklayer	4	tinner	2	silversmith	1
deck hand	4	upholsterer	2	speculator	1
gunsmith	4	watchmaker	2	wagon maker	1
mariner	4	artist	1	wool dyer	1

Source: U.S. Census of Population, 1850.

FIGURE 3.1. *Garten Verein, Galveston's historic German social club and meeting hall, was restored in 1998 after a million-dollar campaign. (Photo: Susan E. Hume)*

on data that compared 1850 manuscript census names with addresses from the Galveston city directory for 1856–57, are shown in maps 3.3 through 3.6.

Many of Galveston's Catholic Germans attended St. Joseph's Catholic Church (fig. 3.2). Others attended the first German Evangelical Lutheran Church, built in 1850. Lutheran churches in Galveston offered services in German until World War I. The Catholic church shown in figure 3.2 was designed by a German architect, Joseph Bleinke, who was one of the German immigrants to Galveston. He arrived in the city in 1850 and helped build this church ten years later.

AFRICAN AMERICANS IN EARLY GALVESTON

Esteban, a Moorish slave traveling with the ill-fated Cabeza de Vaca expedition, was the first African to set foot on Galveston soil, in 1528. Since that time African Americans have continued to settle on the island, first as slaves and later as free residents. Of all the fifty states, Texas has the distinction of having the longest continuous settlement of African Americans.

The census of 1850 lists Galveston's total population as 4,177. Although data on the number of African Americans in the city are unreliable, the

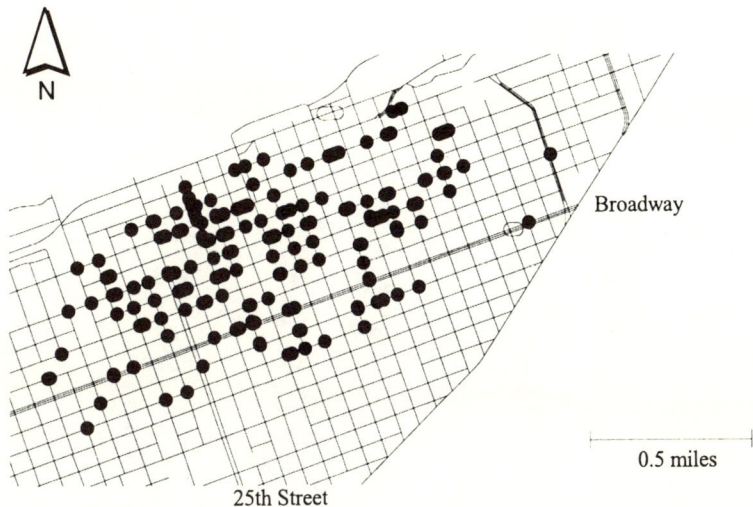

MAP 3.3. *German residential patterns in Galveston, 1857. (Cartography by Linda F. Prosperie)*

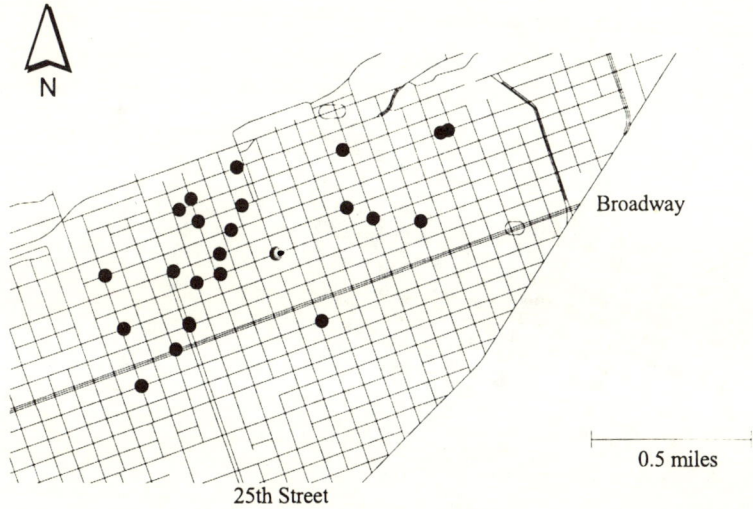

MAP 3.4. *French residential patterns in Galveston, 1857. (Cartography by Linda F. Prosperie)*

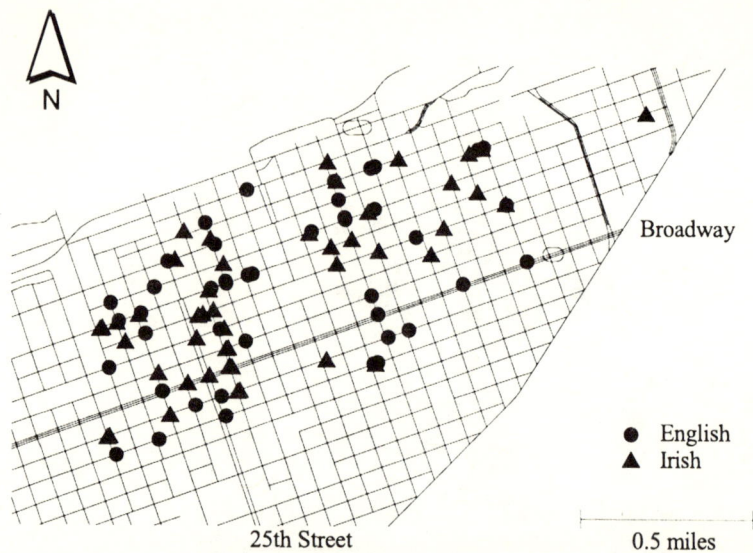

MAP 3.5. *English and Irish residential patterns in Galveston, 1857.*
(Cartography by Linda F. Prosperie)

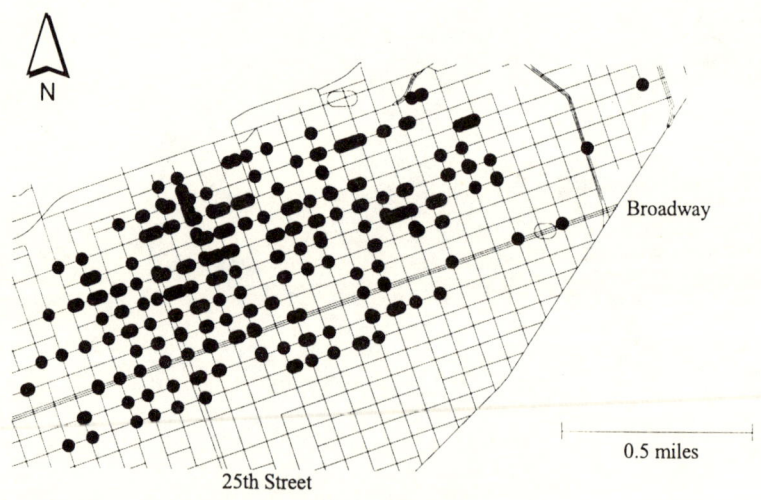

MAP 3.6. *Immigrant residential patterns in Galveston, 1857 (all groups).*
(Cartography by Linda F. Prosperie)

FIGURE 3.2. *Most of Galveston's original German Catholic immigrants attended this German Catholic Church located in a busy residential neighborhood just off Broadway. (Photo: Susan E. Hume)*

TABLE 3.3. Total Slave and Free Populations of African Americans in the United States, Texas, and Galveston, 1850 and 1860

	Slave			Free		
Year	United States	Texas	Galveston	United States	Texas	Galveston
1850	3,204,313	58,161	678	434,495	397	30
1860	3,953,760	182,566	1,178	488,070	355	2

Source: *Negro Population of the United States, 1790–1915*, 1968, 57.

census lists 678 slaves and 30 free African Americans. By 1860 the total population had grown to 7,307, of which 1,178 were slaves. By 1860 the number of free African Americans had dwindled to only two. The manuscript census also lists 261 Galveston slave owners by name; their slaves are listed by age and sex but not by name or residential address. According to Beasley (1996, 18), there were at least 321 slave houses in the city. Most were small bungalows without running water or heat, located along Galveston's crowded alleyways behind the much larger homes of their owners. After the Civil War, these small alley houses became the primary housing available for rental by free African Americans (Beasley 1996, 23). Many continue to be occupied by African Americans in Galveston today. Table 3.3 lists the comparative age and sex of Galveston's slaves and "free colored" residents.

Under the laws of the republic of Texas, free persons of one-eighth "Negro" blood could not vote, own property, testify in court against whites, or intermarry with them. Despite these and other, more local laws against African Americans in Texas, a trickle of in-migration of free African Americans continued, especially into the eastern and southeastern parts of the republic. The city of Houston was especially harsh in its treatment of free African Americans. In this frontier settlement as early as 1839 the City Council adopted a strict city ordinance, which allowed its grand jury to charge free African Americans with being criminals, lazy abolitionists, and bad influences on slaves (Barr 1996, 8).

After Texas was annexed to the United States, even harsher restrictions were adopted, including punishment for minor crimes by branding, whipping, pillorying, and forced labor on public works projects. These cruel restrictions and punishments make it unsurprising that the entire state of Texas recorded a decrease in the number of free people of African Ameri-

BIRTH OF A GLOBAL CITY

can descent from 397 in 1850 to 355 in 1860. In Galveston County, the most populated area of the state, only nine slaves received their freedom between 1840 and 1860. After emancipation, most worked in Galveston as barbers (e.g., Henry Sigler), nurses (e.g., Mary Madison), or worked in livery stables, grocery stores, or saloons.

During the Spanish and Mexican periods of rule in Texas, people of color apparently enjoyed much greater freedom and more economic opportunities. During the earliest years of settlement, Frederick Jackson Turner's concept of the frontier as a "safety-valve" for persons who wished to escape the problems of the settled parts of the country seems to have applied to the African American experience. The institution of slavery barely existed in Texas during the Spanish era. In 1783 the Spanish census of Texas listed thirty-six slave owners in the entire territory, some of whom were arrested by Spanish authorities. As discussed earlier, Frenchmen such as Louis de Aury and, later, Jean Lafitte imported slaves into Galveston between 1816 and 1821. The slaves were sold at an average price of $140 to Louisiana traders such as Jim Bowie for resale across the border.

After Texas became a republic, however, everything changed. Anglo-Americans, mostly of southern origin, adopted harsh restrictions against free African Americans in both rural and urban areas (Barr 1995, 12). Some Anglo colonists described their slaves as "indentured servants" to avoid Mexican antislavery laws. Some slaves were brought to Galveston markets illegally, from the West Indies or directly from Africa, where auctioneers sold them to cotton factors, planters, farmers, and even some urban dwellers (Barr 1996, 16). Slave dealer John Sydnor, a former mayor of the city, operated the largest slave market west of New Orleans in Galveston. Serving as his own auctioneer, Sydnor had a cajoling voice that was known "throughout the state" (Fornell 1961, 115). According to Francis Sheridan, writing in 1839–40, "during the time we lay off in Galveston, two vessels with slaves from Martinique land, their cargoes in security the one below Velasco and the other at Sabine. There are now in Texas 20,000 Blacks and all slaves as the law does permit a free black man residing in the country. Of their treatment . . . the great demand for labor, the immense price it fetches, the poverty and the covetousness of the proprietors, all militate against the port 'nigger,' and I fear his leisure moments are few and his lashes frequent" (Pratt 1954, 89).

Most slaves who were separated from their parents and brothers and sisters in early Galveston never forgot the moment. According to Jeff Hamilton, who was thirteen years old at the time, "I stood on the slaveblock in the blazing sun for at least two hours . . . my legs ached. My hunger had become almost unbearable . . . I was filled with terror and did not know what was to become of me. I had been crying for a long time" (Barr 1996, 17).

Several men of Galveston's earliest power elite—men such as Menard and Borden—operated small plantations on the western end of the island and owned slaves who did most of the labor. Their wives also owned house slaves; according to Eisenhour, the third Mrs. Menard owned six, ranging in age from six years old to thirty-four (1983, 15).

In 1851 a city ordinance began requiring that all free African American residents of Galveston provide a bond that promised good conduct. According to an editorial in one of the local papers three years earlier, "the freest negroes in the county are normal slaves, too much indulged by their owners. Such are the worst pests of this community" (*Civilian and Galveston Gazette*, 1848, 1). This social and financial pressure forced many free African Americans in Galveston to choose a master for protection. For a small sum of money, this master promised care during sickness and old age. In return, the master took responsibility for the good conduct of African American residents. This helps explain why, as mentioned earlier, by 1860 the number of free African Americans in Galveston had decreased to only two. Other strict ordinances forbade African Americans to drink, gamble, participate in disorderly conduct, carry weapons, or not be fully employed, and imposed an eight o'clock curfew.

Another reason for the small number of free African Americans in Galveston before the war was the relative youth of the city. There simply had not yet been time for a larger population of urban African Americans to develop in Galveston, as it had in other southern towns and cities. Moreover, the in-migration and settlement of African Americans in Galveston had actually been prohibited by a set of laws passed when the city was incorporated. These ordinances, like the ones in effect in Houston, even prohibited free blacks from holding dances without the permission of the mayor.

At the end of the Civil War, after federal troops landed on Galveston Island on June 19, 1865, to announce the emancipation of African Americans in Galveston, in Texas, and throughout the United States, conditions changed—but not always for the better. "After the initial moments of joy, fear, searching, and confusion as slavery ended, came a longer period of groping to establish a new status for African-American people" (Barr 1995, 41) as Reconstruction began. Many young men of color flocked into Galveston when the war ended, despite warnings of the military, and settled in an enclave near the location of old Saccarap (*Galveston Daily News*, 5 July 1865, 1). Throughout this postwar period and beyond, free African Americans generally were not warmly welcomed in Galveston, except when wharf workers or household servants were in demand. Federal troops tried to drive freed people out of Galveston and other cities or forced them to cut wood for steamships and perform other jobs for the army. So although African Americans were considered "free" after emancipation (they could now vote and participate in court practices), city ordinances passed before 1865 continued to require that all African American residents of the city be fully employed. The police therefore soon filled local jails with black "vagrants." On one occasion, the sheriff caught black prisoners trying to escape and decided to place them in chains. The African Americans, resisting arrest, threw bricks, so the police opened fire. One prisoner who said that he "would not be ironed by any damned white man" was hit in the head, while another was shot in the body; both men died of their injuries (*Flake's Bulletin*, 20–22 June 1869, as cited in McComb 1986, 87).

Few occupations other than agricultural and domestic work were open to African Americans in Galveston until the early twentieth century. In the decades immediately after the Civil War, African American laborers were seen as a direct threat to white artisans in Galveston. In 1872, expressing an all too common view, a white member of the International Working Man's Association told his fellow members, "If the colored man is to be taken into full fellowship in this society, socially and politically, I must decline to become a member" (Barr 1995, 59). The powerful Screwmen's Benevolent Association of Galveston, formed in 1866, rejected the idea of admitting African American members; black dock workers formed their own organization five years later. (Both these labor unions are discussed in more detail in chapter 4.)

Lack of available capital was one of the major reasons why no African American business class existed in Galveston. Even if this limitation could be overcome by small business owners, such as barbers, other factors, such as lack of patronage by whites and city taxes on small business owners, made things worse. In 1890 the city of Galveston levied a tax of $50 a year on hack drivers, $20 on shoemakers and tailors, $50 for restaurants serving alcoholic beverages, and $5 a chair for each barber in a shop (Rice 1971, 195). For some of these Galvestonians, the fees represented more than two weeks' wages.

THE SHAPING OF PLACE: GALVESTON'S URBAN MORPHOLOGY

People from these two large groups of residents, German and African American, along with English, French, Irish, Scots, and other newcomers, were well aware of the lack of urban amenities on the island. According to several visitors' accounts, Galveston lacked many of the features desired by its residents. It was a low, flat, windswept, and desolate sandy beach that suffered frequent floods from violent storms and was virtually useless for agriculture. The island was solitary and monotonous—as well as a place of disease and danger.

Despite this lack of "creature comforts," by 1839 Galveston had two hotels with three others under construction to house the newly arriving immigrants and visitors to the island. There were also three large warehouses, fifteen retail stores, several lumber yards, six taverns and coffeehouses, two printing offices, drugstores, confectioneries, fruit stores, bakeries, slaughterhouses, and oysterhouses (Wheeler 1968, 71) to support the almost constant stream of new immigrants, along with Galveston's three thousand residents. Oleanders had been planted by new residents in many parts of the city, giving it a colorful aspect still visible in today's landscape. In addition to the improvements created by landscaping along city streets and in residential neighborhoods, by the late 1850s iron fronts had been fitted over brick and frame façades of many of the buildings (fig. 3.3), making the city scene even more attractive to visitors and residents alike (see Fornell 1961, 33–34).

During this same period construction was completed on the first two of Galveston's most substantial homes, both of which are still standing today: Michael Menard's house at 1603 Thirty-third Street (fig. 3.4) and Samuel Williams's Greek Revival–Louisiana Classic residence on a 22-acre lot at 3601 Avenue P (Miller 1983, 62, 66).

Homes in Galveston reflected the varied origins of their owners, and also the heritage of those who built them. Most were two-storied buildings with high ceilings, high windows, wide doors, and broad veranda doors opening from upstairs bedrooms. All homes on the island had many doors and windows to take advantage of island breezes in hot summer months. Many of the houses also had a raised first floor to accommodate potential storm damage and to provide added ventilation (Fornell 1961, 92–93). Behind the houses many homeowners planted thick tropical gardens, blooming southern plants and bushes, and palm trees. Behind the gardens were stables, kitchens, outbuildings for storage, and modest houses for the owners' slaves. (Ellen Beasley's book, *The Alleys and Back Buildings of Galveston* [1996], is highly recommended as a source of additional information on the city's unique alley landscapes.)

Galveston's earliest religious landscape kept pace with its rapidly changing residential and commercial landscape. Roman Catholic services were the only choice of Christian religions until 1840 because no other faith was

FIGURE 3.3. *This Victorian house, like many of Galveston's late nineteenth-century homes, features both iron and brick fronts. (Photo: Susan E. Hume)*

FIGURE 3.4. *The founder of the city of Galveston, French Canadian Michel Menard, had this home built on Thirty-third Street. It is one of the oldest residences on the island. (Photo: Lee Miller)*

permitted by the Mexican government before the Texas Revolution. By 1847 the city became a diocese, which originally included all of Texas (Miller 1983, 71). Father John Odin from France was the city's first bishop. In 1847 a group of Ursuline nuns arrived in Galveston from New Orleans to establish the Ursuline Academy. Work was also started on St. Mary's Catholic Church on Church Street that same year.

The earliest Protestant congregation to organize in Galveston was Presbyterian. This group was founded by a Pennsylvania minister in January 1840 and constructed a church building by the following year. Later that same January the first Baptist congregation was organized, followed by the establishment of a Methodist group. There were quite a few Episcopalians among the first citizens of Galveston as well, and in the early 1840s they formed an Episcopal parish (Morgan n.d., 19–23). By the end of 1845 Galveston had five churches, with two more being built by the Baptists and

German Lutherans. African American slaves organized the African Baptist Church in 1846 and the African Methodist Episcopal Church two years later. The first Jewish congregation in Texas started meeting for services in 1856 in a German member's home, finally completing their synagogue in 1870. By the time the first city directory was published in 1857, thirteen churches had been constructed in Galveston (fig. 3.2). Map 3.7 shows the location of these churches and other features of the cultural landscape of the city as it appeared in 1857.

Educational institutions in Galveston were all privately operated in the years before the Civil War. The City Company built a two-story building, and the state of Texas set aside funds to be used for a public school system in 1854. However, the lack of local tax support and shortage of trained teachers hampered progress in public education. Most teachers came from New England since Texas did not have teacher training colleges until later in the century (*Galveston Daily News*, 1 December 1857, 1).

As in today's school system, it was the students from low-income and often illiterate homes who suffered the most from public education's slow start in Galveston. Wealthy families had the ability to send their children off to boarding school outside of Texas, to hire private tutors, or to send them to private schools run by local Presbyterians, Catholics, or Methodists. The most rigorous and successful private school was the Ursuline Academy for Girls, established in 1847, and a comparable boy's school under the direction of the Oblates of Mary Immaculate (*Galveston Daily News*, 10 November 1857; *Civilian and Galveston Gazette*, 4 August 1857).

Higher education was available on the island, at least for the well-to-do. Young women from wealthy families who wished to stay at home to pursue their college degree could attend either the Galveston Female Collegiate Institution (Presbyterian) or the Catholic University of St. Mary's for women. The curriculum at St. Mary's was impressive. For a yearly fee of $160, students were taught Latin, Greek, and English-language skills, along with mathematics, calculus, geometry, mechanics and astronomy, chemistry, natural philosophy, history, poetry, rhetoric, surveying, botany, and mythology (Fornell 1961, 71).

Along with these educational and religious institutions, social and service organizations helped "civilize" the frontier town. Significantly, however, early membership lists for the Masonic Fraternity, the Independent Order of Odd Fellows, the Samaritans, and Daughters of Samora contain no members with foreign surnames. The Turner Association of Galveston, in contrast, in 1857 had a majority of members with German surnames.

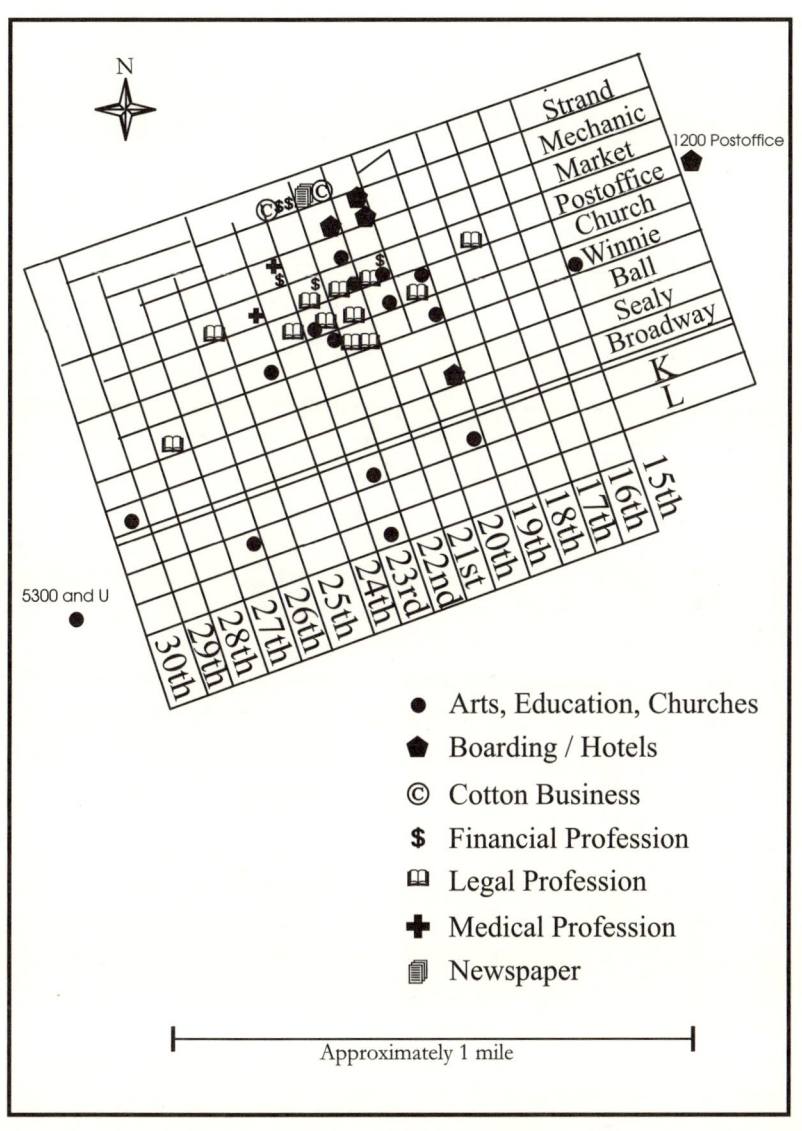

MAP 3.7. *Galveston, 1857: birth of a global city?*
(Cartography by Linda F. Prosperie)

Leadership of this group was also dominated by Germans, who served as "first and second speakers, secretary, assistant secretary, treasurer, professor of gymnastics, and leader of music" (*Galveston City Directory for 1856–57*).

Other groups formed in the earliest years of the city included several women's sewing circles and the Galveston Artillery Company, a social and military organization. In addition to religious, educational, and social institutions, Galveston had a lively arts and literary scene. A German Reading Room was established in 1854 at the corner of Post Office and Center Streets. In addition, a Galveston Lyceum hosted regular debates from 1868 to 1931 and a Galveston Historical Society was founded in 1871. The Tremont Opera House at Market and Tremont opened before the Civil War, and the Grand Opera House, a "thespian temple," opened in 1895.

Economic development was also reflected in the city's morphology in its earliest years. Crude wharfs were first built at the end of Eighteenth, Twentieth, Twenty-first, and Twenty-fourth Streets. In 1836 barely one ship arrived in the port each month. By May 1839 there were often thirty vessels in the harbor at one time (Wheeler 1968, 71). More than two hundred ships visited the port each year. Regular steamship service from Galveston went to Vera Cruz, Tampico, and Havana, as well as to and from New Orleans. Smaller ships crossed the bay and navigated the rivers and bayous to Houston or other mainland villages. A steamer also made daily trips to the island with fresh water from Gulf Coast rivers (Cartwright 1991, 77). This trend toward economic internationalization of the city continued when Texas ratified new commercial treaties with France, England, Germany, and the United States. Cotton, grown on the rich river lands of the mainland just beyond Galveston Bay, was the city's primary export.

A group of wealthy entrepreneurs led by Menard and Williams joined forces to form the Galveston Wharf and Cotton Press Company in 1854. Their effort ultimately merged several wharf companies into one large conglomerate. Because the city owned a third of the stock in this monopoly, the Galveston Wharf Company was considered a semipublic operation and so was not taxed (Wall 1928, 3). Due to this corporation's almost total control over the port, and therefore its ability to raise tariffs and port fees to suit its needs, it was known by many Texans as the "Octopus of the Gulf" through the late 1850s. Many of the laborers working in the cotton-processing centers and the cotton wharves were new immigrants and people of color (figs. 3.5 and 3.6).

According to historian Earl Fornell, the control of almost all port exchanges by the Galveston Wharf Company ultimately had more of an impact on the rise and fall of Galveston than did the Civil War or even the

FIGURE 3.5. *A diverse group of laborers at Galveston's cotton-compressing operation at the warehouses on the docks illustrates the important role of immigrants and African Americans in the city's economic development. (Courtesy of the Rosenberg Library, Galveston, Texas)*

FIGURE 3.6. *Cotton bales stacked at the Galveston Wharf. (Courtesy of the Rosenberg Library, Galveston, Texas)*

issue of slavery (Fornell 1961, 20). Entrepreneurs and investors not associated with this monopoly were forced to look for another suitable site for a port. In the end, this helped launch the move to develop other places farther inland along Buffalo Bayou, thereby "discovering" the potential for development in a tiny village called Houston. According to Fornell, "Galveston entrepreneurs employed their control over the port primarily as a device to build their personal fortunes rather than as a device to develop the island and the adjacent mainland into a great seaport city . . . occupying a geographic site which was threatened by grave risks from ocean storms, the Islanders expanded their facilities with a caution not in keeping with the dynamic aspects of the potential trade prospects then existing in Texas" (1961, 20).

This conservative, controlling style of leadership at Galveston's busy port eventually contributed to the city's economic decline. Because of its protected location and the hopeful prospects of the transcontinental railroad, the developers of Houston's port, in comparison, felt more secure about their protected port and developed policies leading to more dynamic growth, thus outpacing Galveston's economic development by the early twentieth century.

CONCLUSION

The decades before the Civil War were a time of rapid growth and booming economic development on Galveston Island. From a small Mexican outpost with only one permanent building before 1835, the city rapidly evolved into a densely populated place of more than 7,000 residents by 1857. Contrary to popular opinion, however, many of these new residents did not always live in harmony—either with their natural environment or with each other. Sand dunes were leveled and turned into city streets, wharves were built out over the harbor at portside, European and African diseases were introduced, and houses and commercial buildings replaced native vegetation on the increasingly crowded eastern end of the island.

The mid-nineteenth century also witnessed both visible and invisible social stratification of the city's new residents. The physical and human geography of this small island contributed to its evolution into a well-bounded and unique place where a few native-born families controlled most of the money, space, and power. For more than sixty years, the Galveston City Company controlled most of the physical development of the city. This small group of wealthy entrepreneurs known as "The Committee" assumed almost total control over activities at the port of Galveston. After the first municipal charter was granted to the city in 1839, local laws were passed limiting the rights of certain groups of people. African Americans who were not slaves, for example, were forced to register at the mayor's office and were not allowed on the streets after eight in the evening. Free African American residents were also required to pay an additional $75 per month to the city when they rented a house on the island (Wheeler 1968, 73).

This confining and socially stratifying political structure lasted less than a year, but the next city charter limited the rights of certain groups even more. Perhaps the most damaging requirement in the long run was the decision to limit voting rights and tighten requirements for people interested in participating in city government. The original city charter in Galveston allowed all white male property owners to vote, according to Texas law. Several of the Galveston City Company officials wanted more power, however, so they ran one of their own people, Samuel May Williams, as a candidate for the Fourth Texas Congress in 1839. Once he was in power, he pushed a bill through the legislature that limited voting to those white males who owned at least $500 worth of property, thus eliminating half the electorate of the city. Until after the Civil War the city of Galveston did not have universal white male suffrage and, of course, did not even consider the idea of permitting women to vote.

The electoral system and land development effort put control of much of the island's natural and human resources into the hands of a few wealthy families that were mostly native born. These professionals and entrepreneurs, who controlled much of the wharf and city politics, did not represent the real "majority" that made up the bulk of the city's population and dominated both its residential and commercial landscapes. These were the newly arriving immigrants, most of whom came from Germany. The second tier in the social structure of the city consisted of wealthy German entrepreneurs. These individuals held power in their own ethnic circle and interacted regularly in the economic dealings of the community, but they did not mix socially with wealthy, native-born residents.* Below the German business elites were newly arriving immigrants from other European nations. At the very bottom of this tiered system were slaves and, later, free African Americans. African Americans were at the foot of the social and economic ladder,

---

*Despite their social separateness from native-born residents, some Germans in Galveston did amass enough wealth and power to be included in Hayes's prestigious *History of the Island and City*. Biographies of Rosana Osterman, Charles Beissner, Michael Seeligson, F. W. Schmidt, A. Heydecker, Ferdinand Flake, Gustave Ranger, and Sampson Heidenheimer are included (1974, 891-971).

somewhat below Mexican Americans and well below native-born residents and immigrants in Galveston.

As was the case in other southern cities, a complex relationship between white and African American residents of Galveston evolved over the years and included endless groupings and distinctions. After 1865 emancipated slaves generally tried to flee the countryside where they had been living and move to towns and cities (Wade 1964, 249). The position of the free African American in Galveston, as in other cities in the region, was precarious, occupying a middle ground somewhere between slavery and freedom. "We know full well that the pretense of any real freedom being designed or expected for these negroes is but a sham," according to the New Orleans *Picayune* (quoted in Wade 1964, 249). Despite this subservient position in Galveston society, free African Americans tried in every way possible to form support groups by organizing churches, schools (such as the all-black Central High School established in 1885 and discussed in detail in chapter 4), clubs and societies, and community projects. Their presence in nineteenth-century Galveston, therefore, as in the island city today, has played a major role in shaping the city's cultural landscape and social character.

4
Rise of the Elite

On October 29, 1870, the vessel *Heiress* arrived from Liverpool to join the other ships in the busy Galveston harbor. On board the *Heiress* were a thirty-eight-year-old English laborer, Bill Power, and his wife, Christine. Both of these passengers, like so many others on the ship, were undoubtedly tired but excited about their first views of Galveston.

The Powers could not have helped but notice the blue flag with white letters reading "Bethel" flying high over Bean's Wharf as they disembarked. As it did every week, the flag notified the community about a church service to be conducted by the Seaman's Chaplain at three o'clock the next afternoon. Bill and Christine may also have seen the stacks of the *Galveston Daily News* on their way through town. If they had the energy to read through the front-page ads, they might have selected Martin A. Davey's Phoenix Saloon on the corner of Strand and Twenty-fourth for a "good hot lunch and drink to wash it down."

According to the paper that day, the weather was balmy, with temperatures in the mid-70s to low 80s. A Galvestonian, S. N. Moody, had been awarded seven blue ribbons at the "Great State Fair of Texas." And West End residents of the city were told that they were "on the verge of having an excellent, good hard road to their residences" built. According to the front-page feature story, this new road would be the direct result of Mayor Leonard's decision to tear down the old Magruder Breastworks so that Avenue O could be extended all the way out to Tremont Street. The paper also reported that the Galveston Recorder's Court had fined a man a dollar for "making an ineffectual attempt to cut a man's head off with an axe." Strangely enough, in the mores typical of a frontier town, a young woman had been fined the same amount for "street walking."

Two years later, Bill and Christine had found a small house near other English immigrants on Twentieth Street between J and K. Bill worked as a salesman at Klopman and Fellman's, a small dry-goods store located on the corner of Twenty-first Street and Avenue D, only six blocks from the Power residence.

By 1875, according to listings in the Galveston city directory, Mr. William Power had become a co-owner in the thriving dry-goods business. The original owners had moved back to New York, leaving Bill in charge of the store's dry goods and groceries. A visit to the Power house more than a hundred twenty-five years later reveals a small, tidy frame home just off Broadway. The original site of Bill's Klopman and Fellman Store is now a vacant parking lot bordered by the drive-through window of one of Galveston's downtown banks.

The Powers' experiences in post–Civil War Galveston were typical of those of many immigrants. They left home to find a better life. They expected to find improved economic opportunities in what sounded like a more exciting place, and that is exactly what they found. What they did not expect was the power money held over most of the city's residents. Like other immigrants living in the city, they soon learned that money and power were inexorably linked—and both governed who made most of the decisions shaping the island's future.

In the years following the Civil War up to the Great Storm in 1900, Galveston experienced rapid and continuous growth, ultimately developing into what one writer called the "Wall Street of the Southwest." According to popular publications of the day, extolling the city's location and potential for continuous growth—along with the opinions of business interests in Galveston and beyond—the city's growth and prosperity were destined to last forever. Galveston remained the largest city in Texas up to the census of 1890; the pace of Galveston's trade, the beauty of its architecture, the success of its businesses and industries, and its cosmopolitan ambiance all pointed to a bright future. In the words of one writer of the day, "it was ordained to be the Seaport of the West, with a destiny of maritime ascendancy, of grandeur, and of power" (Morrison 1890, 4).

Much of the high-powered growth of Galveston came about because of its geographical position. According to a description written in the mid-1870s, "Galveston is about 400 miles from New Orleans, 700 from Vera Cruz, and 800 from Cuba . . . with a harbor free of obstruction, the broad ocean before her as the highway to home and European markets and territory equal to an empire, and the lone star state at its crest. The city, the natural outlet for her products, and in which the state feels a pride, will cover the island like New York that of Manhattan Island, and be the great emporium of exports and imports of not only her own great state, but also

RISE OF THE ELITE

of the North and West" (*Fayman and Reilly's Galveston City Directory for 1875-6*, 6).

Even casual observers believed that the city simply could not fail. In 1874 the *New York Herald* honored Galveston as "the New York of the Gulf" (as quoted in the *Galveston Daily News*, 10 March 1874, 1), and locally the city was said to be "like a young giant and the center of all business in the state ... where there is more doing than elsewhere" (Bragg 1874).

Galveston's insular location and limited geographical space, however, would both ultimately prove to be major constraints for ongoing development. The city's image as a sophisticated, cosmopolitan, and wealthy place held true only for a select number of white, native-born residents. Life for the newly freed slaves and poor white residents—most of whom had arrived from the struggle of their homes in the South after the Civil War—continued in stark contrast, as did the economic struggle of many of the city's new immigrants.

CONNECTIONS WITHIN AND WITHOUT

Galveston's location on the edge of the continent, called North America's "Third Coast" today, was all-important for its growth potential. The city's increasingly important role as the pivot of a global transportation system that connected Texas with the interior states of the United States also meant it was connected to people, ports, and products in many other parts of the world. The establishment of effective transportation systems that linked the port of Galveston with cotton and sugar plantations, as well as with people and products of towns and cities in the interior of the state and continent, was essential to its long-term survival.

Early Road and Water Connections

Galveston's direct link to the sea was an early transportation advantage that gave it an edge over other places in Texas. On land, wagon and cart transportation by oxen and horses worked well in dry weather but could be shut down for days or even weeks at a time by bad weather. Interior waterways were also unpredictable because of the wide disparities in seasonal flows of the rivers. The unnavigability of many of these Texas rivers at certain times of the year made water transportation a seasonal facility at best (*DeBow's Review* 23 [1867]:113–26; Richardson 1862, 118). As early as 1850, the Galveston and Brazos Navigation Company began to construct a canal between the harbor of San Luis in West Galveston Bay and the Brazos River. This new waterway, four and a quarter miles long, was completed five years later, linking Galveston Island with the mainland. Even though the canal could accommodate both sailing vessels and steamships, in the long run it could not compete with railroads and river transportation (Webb 1952, 664).

Supporters of the construction of plank roads, along with canal and riverboat entrepreneurs and railroad promoters, all claimed to have the answer to the state's transportation problem (Fornell 1961, 179). Local developers and state planning agencies alike soon agreed that plank roads and railroads were the only way to effectively link the city of Galveston with the interior throughout the year, even in bad weather.

Railroads

The ever intensifying power struggle between the cities of Galveston and Houston (discussed in more detail in chapter 5) first centered on decisions about the construction of state-linked railroad lines. After a decade or more of debate and disagreement about where to build all-important railroads in Texas and who would finance their construction, by 1860 about four hundred miles of track had been planned for the state's railroad system, including lines for the Houston and Texas Central, the Texas and New Orleans Line, and the Galveston and Houston Railroad. If completed, these plans meant that passengers would be able to ride the trains from Galveston to places such as Victoria, Brenham, Houston, Hempstead, Wharton, Richmond, Beaumont, and Orange. Not only would it be possible for new Texas immigrants to settle more easily at places far distant from Galveston harbor, but also it would become feasible for cotton and other products of the city's fertile hinterland to be moved in and out of the port of Galveston, thereby connecting Galveston to the all-important New England and European textile industry (Fornell 1961, 159). Much of the urgency to build railroads came from wealthy agricultural investors on the Texas Gulf Coast and the nearby interior region, as they developed fertile cotton and sugar land drained by the Brazos and Trinity Rivers.

Several plans were developed beginning in the 1840s that suggested optimal routes for Texas railroad development. The early Transcontinental Plan favored linking the state's railroads to the larger national network. It also suggested that the city of Houston, rather than Galveston, should become the railroad hub of the southeastern part of the state. Galvestonians objected. They felt that the Transcontinental Plan should be replaced by the Galveston Plan because "the logic of geography dictated that all Texas rail lines must lead 'fan-like' to Galveston" (Fornell 1961, 16). The editor of the Galveston *Richardson's News*, Willard Richardson, pointed out repeatedly in his front-page stories that building railroads according to the original Transcontinental Plan would make Texas dependent on either New Orleans or St. Louis and so should be avoided.

As discussions and decisions about railroad routes in Texas continued, it became increasingly evident that the Houston-transcontinental viewpoint would dominate at the state legislative level. Even Galveston's newspaper editor Richardson finally conceded that the Houston-centered Transcontinental Plan no doubt would win when the votes were counted in Austin. He thereafter decided to join the cause, but strongly endorsed the plan to have all lines built and controlled by the state of Texas (rather than owned by private capital provided primarily by wealthy Galveston investors). Enthusiastic support for this state-based plan also came from two other large groups—the lower- and middle-class citizens of Galveston and small farmers in the interior. Both groups distrusted private-sector promoters. They also supported the plan that would develop and activate railroad operation as quickly as possible (Fornell 1961, 166). Despite the preference of Galveston's lower- and middle-class citizens, however, a bill was passed in the Texas legislature in 1856 that specified that all railroad lines in Texas were to be funded and built by private corporations (Richardson 1862, 132–34).

Soon after this decision, a group of wealthy Texas promoters—owners of the Buffalo Bayou, Brazos, and Colorado Railway Company—launched their plan to construct the first rail line in the state. Its primary role would be to transport cotton and sugar to Galveston Bay. This rail line, completed in 1876 as a narrow-gauge line 15 miles long, would connect Galveston with a few small towns in the interior (Webb 1952, 664). This line was abandoned in 1880, however, because it failed to link the port of Galveston with planters and their products in places such as Brazoria, Matagorda, and Wharton—all important areas of cotton and sugar production. Plantation owners in this interior region of the state ultimately built their own line, known as the Sugar Railroad, which extended north from Columbia to connect to the Buffalo Bayou, Brazos, and Colorado Line (*Civilian and Galveston Gazette*, 22 March 1859, 1).

In 1870 the Sugar Line had its name changed to the Galveston, Harrisburg, and San Antonio Railway Company and connected the town of Alleyton on the Colorado River with the city of San Antonio. Its route was enlarged again in 1884 to include consolidation with five other lines: the Gulf Railroad, Western Texas, and Pacific Railroad; the New York, Texas and

Mexico Railroad; the San Antonio and Gulf Railroad; the Gonzales Tap Railroad; and the Galveston, Houston and Northern Railroad. This expansion meant the line now had port connections at Indianola and Fort Lavaca as well as at Galveston. In 1905 the entire network of railroads forming this consolidated system became a part of the Southern Pacific transcontinental system (Webb 1952, 665).

Meanwhile, Houston's plans to become the central rail location in southeast Texas remained a matter of great concern to many Galvestonians. Their fears materialized when the Texas and New Orleans Railroad was approved by most of the Texas legislators and constructed with funding supplied by New Orleans promoters. Galveston was not even considered in this plan. These promoters failed to deliver their support, however, so this Houston-centered railroad was never completed.

Replacing the defunct Texas and New Orleans Railroad as a link between Galveston, Houston, and New Orleans was the Galveston, Houston, and Henderson Line, which linked the two southeast Texas cities through the Civil War years. This important railroad crossed Galveston Bay via a bridge from the mainland after 1860, making it possible for supplies to be transported between the island and the mainland. The all-important function of this line seemed to ensure Galveston's central role as the state's main port of entry.

Completion of the causeway bridge for the Galveston, Houston, and Henderson Railroad provided the island with its first direct link with the mainland. This railroad reached as far north as Houston by 1860 but never extended any farther north because of the disruption caused by the Civil War. Financial backers from Houston eventually withdrew their support, fearing that all shipments from the interior would pass directly through Houston and be unloaded at Galveston.

In the end, as a break-of-bulk point, Houston had the advantage over Galveston All cargo brought in or out of this more interior city had to be transferred onto another rail line to get to Galveston. This expense, along with yellow fever quarantines imposed on Galveston shipments at fall cotton harvest, encouraged growers to ship their cotton in and out of Houston directly rather than take on the added trouble, time, and expense of sending it on to Galveston (Scheibe 1992, 42). In an interesting aside, it is recorded in numerous places in Galveston newspapers and other public documents that Houston officials were known to often fake the news about outbreaks of yellow fever in an attempt to aid their cause—keeping Galveston from receiving timely shipments.

Meanwhile, business interests in Galveston were desperate to build a

RISE OF THE ELITE

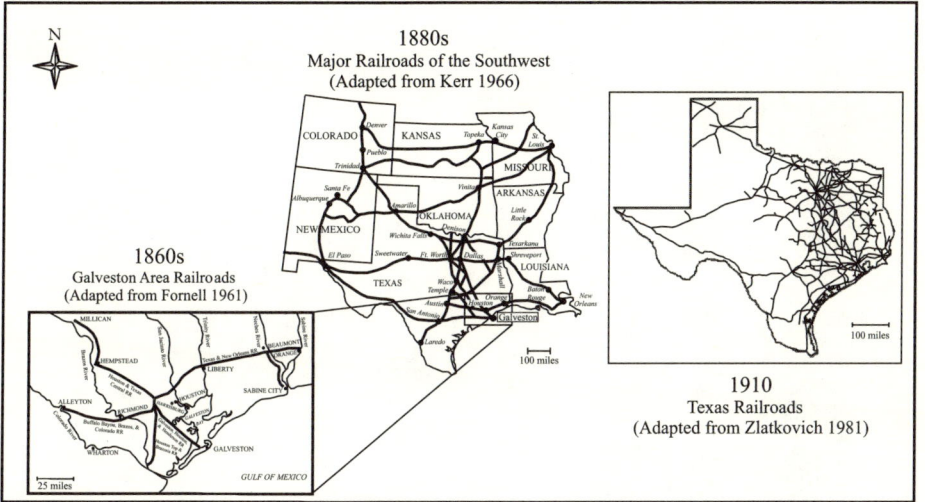

MAP 4.1. *Railroad connections to the interior. (Cartography by Linda F. Prosperie)*

railroad that bypassed Houston completely to connect Galveston with the interior of Texas. In 1873 several key Galveston investors chartered the Gulf, Colorado, and Santa Fe Railway Company. Later, in 1890, the Galveston, Houston, and Northern Railway, consolidated from several other lines, was chartered. In 1899 the Southern Pacific System purchased all prior lines and extended them along the Galveston Bay shore. This branch was abandoned in the early 1930s, leaving only 53 miles of mainline track connecting Houston to Galveston today (Webb 1952, 666). Map 4.1 shows railroads that were central to the economic development of Galveston.

Galveston was the first city in Texas to be served by a street railway, started and actively used after 1866. Buses took over all the lines formerly served by this railway in 1938. Up until 1891, when electric cars were purchased and used, Galveston's streetcars were powered by mules. One of the most surprising features on maps of the developed beach district of Galveston in this period is the beautiful wooden pavilion at Twenty-first and Q, constructed on the beach to mark the end of the line of the Galveston Railway Company. Rail, and later the city's streetcar lines, were laid directly in the water (fig. 4.1).

The Galveston city directory listed railroad stations and the addresses of their depots in 1899: the Union Passenger Station at the corner of Twenty-fifth Street and Strand; the Galveston-Brazos-Southwestern Railroad with a depot at Tremont and Water Streets; the Galveston-Houston and Hen-

FIGURE 4.1. *Galveston's streetcar line ran right out onto the beach on the island's Gulfside until the Great Storm of 1900 destroyed the tracks. (Courtesy of the Rosenberg Library, Galveston, Texas)*

derson Railroad at the northwest corner of Rosenberg and Twenty-fifth Street; the Galveston-Houston and Northern Railroad, also headquartered at the corner of Rosenberg and Twenty-fifth Street; the Galveston and Western Railroad at 2209 Post Office Street; the Gulf-Colorado and Santa Fe Railroad at Rosenberg and Twenty-fifth; and the Gulf and Interstate Railroad with its depot on Pier 18. The same directory also listed a total of ten streetcar lines with seventy-three cars running, operated by the Galveston City Railroad (*Morrison and Fourmy's General Directory . . . 1901–1902*). The route of the Galveston to Houston Interurban Railway is shown in figure 4.2.

Impacts of the Civil War on Transportation

Galveston's strategic location as the main port on the Texas Gulf Coast made it especially vulnerable to attack during the Civil War. As a defense against the Union troops, Fort Point was built on the northeastern point of the island in 1861. Despite the construction of these fortifications along

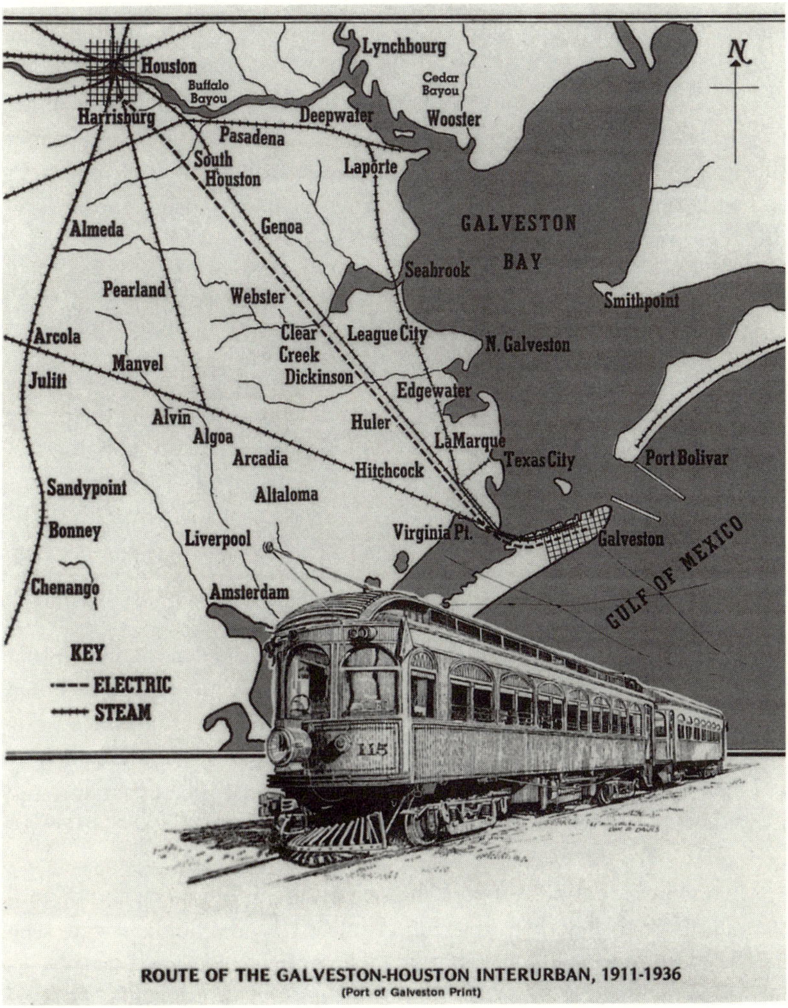

FIGURE 4.2. *Galveston was an important hub for the Galveston–Houston Interurban Railroad between 1911 and 1936. (Courtesy of the Rosenberg Library, Galveston, Texas)*

the outer edge of the island, however, overall defense of the island remained weak.

The Union forces threatened to shell Galveston Island in the spring of 1862, but after long discussions with their superiors in Washington and foreign consulates in the city, they agreed to put a complete embargo on the port instead (*Tri-Weekly News,* 23 May 1862). Union military authorities, however, still worried that the Confederate forces would attack to break down the embargo. These fears led Union leaders to ask local laborers to remove all coal and other products from the port of Galveston. Despite offers of high pay, the mostly Confederate local laborers refused to cooperate. In consequence, the Union military authorities decided to put Galveston under martial law. Surprisingly, most Galveston residents were relieved to hear of this decision because they believed it would protect their city from attack and potential destruction. The *Tri-Weekly Telegraph* published the news on the front page of its May 23, 1862, issue: "Martial Law—We publish today General Order No. 41, of Gen. Hebert, declaring Martial Law. We are glad to see this, and have hoped for it for some time. It will work well we have no doubt."

As predicted, issuance of this general order brought calm to the island. But all was still not well, for all Galveston male residents between the ages of eighteen and fifty-five were forced to enlist in the Union army. The only men who were excused from this conscript law were new immigrants who were able to secure official protection papers from the consulates of their various countries. This intensified bad feelings between native-born residents of the city and Galveston's immigrant community. As further evidence of the multicultural character of Galveston in the early 1860s, the same newspaper reported, "We learn from the Union that 298 persons claiming citizenship and foreign protection, have recorded their names in the office of the Provost Marshal in Galveston, since May: 133 Germans, 43 Englishmen, 62 Frenchmen, 5 Spaniards, 37 Portuguese, 2 Italians, 7 Danes, 2 Belgians, 3 Swiss, 1 Dutchman, 2 Hungarians, and 1 Swede. These represent nearly one-fourth the population of Galveston" (*Tri-Weekly Telegraph,* 28 July 1862, 1).

Union forces continued to stop ships entering the Galveston harbor until the fall of 1862; they then took full claim of the city. Soldiers were seen on the island during the day, but at night military personnel usually returned to fortified positions on their ships docked around Kuhn's Wharf on the edge of the port. This situation was not to last for long. In what is now remembered as an exciting and typically Texan middle-of-the-night battle in January 1863, the military leader of the Confederate forces, General

John Magruder, and his Galveston-based troops, captured over five hundred Union men. These prisoners were shipped to Houston on the newly constructed Galveston, Houston, and Henderson Railroad. After this battle, the North's embargo on the port continued, but until after the war the city of Galveston was never again occupied by Union troops. By the end of the war the North had occupied every port in the South except two—Charleston and Galveston.

The Civil War played a major role in slowing Galveston's growth and development, not only by decreasing the influx of new residents but also by forcing many of its citizens to leave for safer ground in the Texas interior. Witnesses recorded that a great deal of destruction to the built environment of the city was evident when the war ended. Fences, homes, barns, and outbuildings had been ripped apart for firewood, army defensive mounds still ringed the town, and cannonballs were lodged in the walls of homes and buildings throughout the city (Eisenhour 1983, 19). There can be no doubt that four years under the constant threat of destruction by both the Confederate and Federal forces had brought the city to the verge of complete deterioration by the end of the war.

But Galveston quickly rebounded. The rapid rehabilitation of the city's business firms, port facilities, and residential, commercial, and industrial buildings moved forward efficiently after 1865. Shipping lines reestablished contact between the port's cotton markets and the outside world, as business interests on the island continued to look for more direct connections to the interior.

The end of the war also brought about a more bounded and more clearly visible social territory dividing certain ethnic and racial groups in the city. This was especially true in the first two decades after the war. In these years, newly freed slaves were generally allowed to rent only the former slave cabins on alleyways behind the city's larger, white-owned homes. This division, with whites in the front of lots and African Americans in the rear, helped shape long-term residential patterns in the city. These firmly established location patterns became difficult, if not impossible, to alter even many decades later.

Steamships

After 1865 steamship lines connected Galveston with New Orleans, New York City, and Western Europe. At the end of the war, two of the lines were dominant—Spofford, Tileston, and Company, and Williams and Guion (Scheibe 1992, 43). Steamships also operated regularly between Galveston and a variety of European ports, bringing immigrants into the city from countries such as Germany, France, Russia, and Spain. In the spring of 1868 the steamship *Pioneer* left Galveston for Liverpool carrying the first load of Texas cotton bound for the European textile industry. Thereafter, the cities of Liverpool and Bremen served as the two most important places of European contact for Galveston shipping lines.

After 1870 two different lines became all-important in the city's ongoing maritime development: the Morgan Line and the Mallory Line. The Morgan Line carried most of the intra-Gulf trade before and immediately after the Civil War. Charles Morgan, the founder of the company, made the all-important decision to invest heavily in railroad development. He offered cotton and sugar growers as well as passengers a combined rail and water transportation system. He later played a pivotal role in the development of the Houston Direct Navigation Company. Through this business alliance, he created a trade environment that greatly benefited his own company while furthering the city of Houston's growth. Morgan joined forces with the newly established U.S. Army Corps of Engineers to dredge a 12-foot channel from the bay to a point 12 miles (6 land miles) from Houston on the edge of Buffalo Bayou. Because of his heavy investments, Morgan and his company were given complete control of the Houston Ship Channel Company, which included not only the right of free navigation for the Morgan fleet but also the opportunity to link his expanding railroad empire with other railroads that served the Texas interior. Soon thereafter, Morgan built a rail line connecting the terminus of the channel with other railroads already serving the city of Houston (Sibley 1968, 96–106). Many Galvestonians still blame Morgan for their city's eventual decline, because of his "defection" from the Galveston port in order to support Houston.

But it would take more than careful railroad planning and construction to save the primacy of the port of Galveston. Because of numerous rail connections between Texas cotton areas and the East, and the invention and proliferation of the use of the telegraph, cotton markets could be established virtually anywhere in the Cotton Belt.

> Intense competition between the various overland transport alternatives placed a vast area of Texas on a "common point or equal rate basis" to the benefit of non-Texas merchants and manufacturers. Such maneuverings greatly aided out-of-state interests in the attempts to wrest control of the general trade of the state from out of the hands of both Houston and Galveston, which, on the basis of proximity alone, might have expected to maintain a virtual monopoly of the trade of much of North and East

RISE OF THE ELITE

Texas. . . . Instead . . . Texas commerce was diverted northward though Denison and Texarkana to St. Louis, Chicago, Louisville, and Cincinnati. . . . This loss in export traffic through the port of Galveston was equaled by the inroads made into the Island's import trade; the railroads carried westward shipments of breadstuffs, dry goods, clothing, boots, shoes, saddlery, agricultural implements, and all kinds of general merchandise. Galveston could not afford to ignore so revolutionary a trend in Texas commerce. To offset these disadvantages of rail connections with the nation, the Island was compelled to look toward the betterment of its waterborne connections. (Kelly 1975, 89-91)

Morgan's efforts to work closely with Houston's business interests caused him eventually to withdraw his fleet entirely from the Galveston harbor. His decisions also added fuel to the fire of competition in the city of Galveston, as plans were made to improve Galveston's deep-water harbor. In addition, this "defection" of the Morgan shipping interests in Galveston opened the door for the Mallory Line to play a dominant role. The New England-based Mallory firm had been started in Mystic, Connecticut, by Charles Mallory, but its headquarters were moved to New York when the founder's son took over. As a result of the move, the company's attention shifted to making connections between East Coast and Gulf of Mexico port cities. By the mid-1860s Galveston had become one of the two most active ports for Mallory. By the end of the decade Mallory had become the dominant line seen in Galveston Bay.

One of the least known migration stories connected to the Mallory ships concerns their role in transporting numerous residents of the southern states who wanted to leave the United States after the loss of the Civil War and relocate in Latin America. These out-migrants all passed through the port of Galveston. Many, such as Galveston's General Magruder and other commanders of the Confederate forces, were drawn to Latin America because they preferred living outside their country of birth rather than to stay under Yankee rule. Magruder later joined the imperial forces of Maximilian, emperor of Mexico. After the defeat of this military leader, Magruder returned to Texas, where he was a popular lecturer on his Mexican military escapades.

After a new steamship line was opened between Galveston and the Tuxpan Valley in Mexico, a new colony of expatriates settled there. This movement was supported by the publication of Major John Henry Brown's popular book *Two Years in Mexico, or The Emigrant's Guide*, especially written for southerners who wished to move to the Tuxpan Valley. In 1867 the editor of the *Galveston News* reacted to this guidebook by lamenting that "the causes which are forcing people to pull up stakes and leave the best government and country the sun ever shown on, for inhospitable Mexico, are subjects of the saddest reflection (*Galveston Daily News*, 26 June 1867, 1).

POSTWAR RECOVERY IN GALVESTON

Two years after the war ended, it seemed that in Galveston all was lost. More than 8,000 people in the city were infected with the dreaded disease of yellow fever in 1867 and 1,171 people died. The city was terrorized by the epidemic (Beasley 1996, 26). According to Tim Finn, a house painter who was a native of Boston, "everything about Galveston had the appearance of a grave Yard no Buissnis of any Kind but to Attend to the Sick and Bury the dead" (Beasley 1996, 26, from the *Galveston Daily News*, 4 July 1865].

Adding to the misery of what was to become the last major yellow fever epidemic in the city was an October hurricane that destroyed several buildings and killed three people. But strangely enough, archival records indicate that more than two thousand new homes were constructed that same year, the city's first gas street lights were installed, and improvements were made at the port. The effort to rebuild the city after the damage of the war years was thwarted again in 1869 when a fire destroyed more than one hundred frame buildings on the Strand, in the heart of the city's commercial district (Miller 1983, 103–4).

Despite numerous setbacks, the census of 1870 recorded a total population of 13,818 in Galveston.* That same year a Polish Jewish businessman, Harris Kempner, moved to the island to launch one of Galveston's most important family-based commercial dynasties. At the same time, a young African American man in his twenties, Norris Wright Cuney, began studying law books in an attorney's office in the city. A few years later he entered the law profession himself. Cuney became a major player in the state Republican party in the 1880s and 1890s, rising to a position of economic and political power that few people of color found it possible to accomplish in the first decades of emancipation. Map 4.2 provides further evidence of the success of Galveston's effort to rebuild in the early 1870s.

As if built to be a symbol for the city's growth, a modern-looking steel

---

*Note that the 1870 census is deeply flawed by undercounting. This deficiency was caused by the Confederates' general unwillingness to cooperate with the Reconstruction government in Texas and much of the South.

MAP 4.2. *Bird's-eye view of the city of Galveston, 1871*

lighthouse tower was constructed by the Coast Guard in 1872 on the tip of Point Bolivar. This tower was located just across the bay from the city's port and commercial district, and it lit the way for ships entering and leaving the harbor until well into the 1930s. New churches, schools, and hotels were also constructed in the 1870s, most still built entirely of wood, but more and more constructed with stone and brick in an effort to protect the city against any more serious fires. The all-powerful Galveston Cotton Exchange was formed in 1872. Galveston citizen A. H. Belo, later to be known as the founder of the *Dallas Morning News,* installed the first telephone line in the state of Texas between his home on Avenue K and his office on Market Street, and the first telephone exchange began to operate officially in the city in 1879. The Brush Electric Company began offering electric service to homes in Galveston in 1882 after the Electric Pavilion on the beach, located at the site of today's Moody Convention Center, advertised electricity as the power of the future. By the time Mardi Gras began in February 1883, a few streetlights had even been installed in downtown Galveston. By 1880 the U.S. census listed 22,248 people living in Galveston, which meant that it was still the largest city in Texas.

SETTLEMENT PATTERNS AND RELATED
SOCIAL PROCESSES, 1865-1900

Much of Galveston's cultural and economic recovery after the Civil War was the direct result of the influx of newly freed African Americans and new immigrants. A surprising number of people also came from the northern states, part of the renewed wave of Americans moving to the West and the South—a migration stream described by the *Galveston Weekly News* as "irrepressible as a tidal wave." These migrants were attracted by the promise of future prosperity in the West and South. In a letter of support encouraging Yankees to move south to the Gulf city, Galveston realtor Henry M. Truehart, wrote, "I know of no reason why a Union man could not live here in perfect safety. I believe the very large majority of the people of Texas would welcome immigrants from any and every part of the United states and be delighted to see them settle here" (Truehart Papers).

Other new residents of the island originated in Europe, some coming from Germany, Great Britain, and Scandinavia as before the war, and others from the nations of southern and eastern Europe where economic and political conditions had deteriorated in the second half of the nineteenth century. The Irish came into Galveston on trains from the north in the years after the Civil War, former laborers who had helped build the rail system that now linked Galveston with the mainland. Many, perhaps, sensed the economic potential of the Galveston area and were drawn to the city by the presence of the small community of Irish residents who had settled on the island after serving at San Jacinto. These later arrivals originated from the Irish counties of Mayo, Galway, Kerry, Cork, Sligo, Leitrim, Cayan, Queens, Westmeath, Tipperary, Dublin, and Down; a few of these names appear as place names on the map of Texas as reminders of the important role the Irish played in helping to settle the state (Flannery 1995, 122). One of the Irish immigrants who played a major role in the urban morphology and design of the city of Galveston was Nicholas Clayton, who was born in County Cork. Because he designed the Bishop's Palace and numerous other architectural relics still standing in Galveston, Clayton is remembered today as the individual most responsible for the architectural ambiance of the city (fig. 4.3).

The Irish immigrants formed fraternal, social, and educational groups in Galveston, as they did in other parts of the state. In 1874 the first Ancient Order of the Hibernians (AOH) division was organized in Galveston, no doubt because "the city's populace included Hibernians who had moved from Boston and Philadelphia areas, and most of the dockworkers were of

FIGURE 4.3. *Bishop's Palace, located across the street from Sacred Heart Church on Broadway, is a popular tourist destination for visitors interested in historic preservation. (Photo: Lee Miller)*

Irish birth" (Flannery 1995, 141). The AOH provided housing and aid for new immigrants and preserved new arrivals' connections with the ancient history and culture of their ancestors. Minutes from some of the first meetings of this organization record discussions of ancient and modern Irish history as well as Irish customs, traditions, and languages.

The Irish and other European newcomers arrived in Galveston aboard steamship lines from places like New Orleans, Liverpool, and Bremen, as well as on railroads from the north. The *Galveston Community Book* (Graham and Newman 1945, 36) describes the impact of European immigrants:

> Down the gangplank at the Galveston wharves have trudged feet from all parts of Europe for over a hundred years, eager feet and tired feet and patient feet and hopeful feet searching for homes, ready to make farms from fertile acres, raise cattle on ranches, build ships, or develop a hundred little businesses. They were bound for many different parts of Texas but those who lingered on Galveston Island have added varied influences of Europe to the culture of the Old South, the Spanish-Indian tradition, the transplanted eastern industry, and the southwestern freedom. The result has been the cosmopolitan tone of Galveston—unique in Texas.

FIGURE 4.4. *Pelican Island, a stone's throw from Galveston Island on the bayside, provided an ideal location for immigration-processing stations like this one. (Courtesy of the Rosenberg Library, Galveston, Texas)*

Although immigrants from many parts of Europe continued to arrive at the port of Galveston, the pace of in-migration changed in the decades after the Civil War. Federal laws in 1875 ended the unrestricted entry of immigrants into the United States. This led to a reduction in the number of immigrants entering Galveston harbor in the late nineteenth century, but certain groups, such as Italians and Eastern Europeans, found their way to the first official U.S. immigration station at Pier 29 in Galveston where local doctors gave medical examinations to newcomers. Customs officials inspected baggage and processed paperwork. Anyone found to be diseased or incapacitated was sent home. This processing center was later replaced by several other immigration receiving and quarantine stations. The last was one of two processing centers located on Pelican Island (fig. 4.4).

The vast majority of Italian immigrants, who came primarily from Sicily, arrived after the turn of the century; their experience and settlement patterns are discussed in detail in chapter 5. Eastern European immigrants also arrived in significant numbers; many were Polish, Hungarian, and Czech families who left almost immediately for the rich farmland and smaller cities in the state's interior. From a geographer's perspective, one of the

RISE OF THE ELITE

most interesting of these Eastern European immigrants was a Pole named William Sandusky who was hired as a draftsman responsible for creating maps of Galveston harbor. Map 1.2 in chapter 1 is one of Sandusky's best known and most often reproduced cartographic accomplishments.

The first large group of Polish immigrants arrived at the wooden Merchant's Wharf in Galveston harbor on the sailing ship *Weser* in 1854. These 150 Polish farmers who had originated in southwestern Poland in a region known as Upper Silesia founded Panna Maria, the first Polish settlement in North America, in central Texas.

A series of environmental, economic, and political factors caused people from Poland to resettle in Texas. First, Upper Silesia had severe economic problems—primarily inflation and a depressed economy—caused in part by the impact of the Crimean War. The region had also experienced severe epidemics of typhus and cholera, and numerous devastating floods. Second, some Poles left Silesia to escape a harsh system of conscription into the Prussian army. Third, Poles faced discrimination by the Germans who dominated politics and economic power in their homeland (see Baker 1982, 10-13). The Poles' failure to gain the advantage in the revolution of 1848 left the Germans holding power over the Poles. Much like the experience of the Wendish people of Eastern Germany, in one of the twists of fate so common in migration stories, Polish emigrants who left their own country found themselves arriving at a port city in Texas in an immigrant community that was also dominated by Germans.

The organizational and motivational skills of a young priest in Poland, Reverend Leopold Moczygemba, along with letters home describing the perceived bliss of life in Texas, encouraged more and more Polish immigrants to migrate to Texas. Father Moczygemba dreamed of becoming a missionary his entire life, and finally in the summer of 1852 he arrived in Galveston. He wrote letters praising life in Texas to his friends and relatives in Upper Silesia, urging them to come to Texas. Like other groups of immigrants from Europe, most traveled by train from their home area to their port of departure. Railroad agents offered special fares and baggage allowances to encourage travel by this new method of transportation. Most Poles traveled directly to Bremen. A few went to other ports, such as Hamburg and Szczecin, sailing on a two-month voyage for either Galveston or Indiananola. Since the majority of early Polish immigrants who settled in Texas were rural, agricultural people, they almost immediately left the port of Galveston to travel inland to settlements like Panna Maria, located just south of San Antonio. Others settled in Bandera, just north of San Antonio. A few, possessing skills useful in a more urban setting, stayed behind in Galveston, no doubt enticed by their first view of this very European-looking city. Other, more urban Polish immigrants eventually formed a recognizable Polish Quarter in the city of San Antonio.

Czechs are another group of Eastern European immigrants who arrived in Texas at the port of Galveston with plans to settle inland areas. Originating primarily in Moravia and Bohemia, ethnic Czechs came from the multinational, and often quite oppressive, Austro-Hungarian Empire. When waves of revolution severely affected their lives in 1848, many gave up hope of ever claiming power in their homeland and so began to look to North American resettlement as their only hope for political freedom and improved economic opportunities. Most of the Czech migrants traveling to Texas were farmers who were drawn to the Gulf coastal plain and the state's blackland prairie region farther north in their earliest years of settlement.

Some Czech migrants did stay in Galveston, however. One such was Moritz Kopperl from Moravia, a prominent banker and businessman in the city. Arriving in Galveston in 1857, Kopperl lost all of his investments during the Civil War but managed to pay back his debts after the war and make a fresh start. In 1865 this enterprising Czech immigrant entered the cotton commission business and, later, the coffee trade (Hewitt 1975, 9).

Because of the severe displacement caused by the revolutions of 1848, which attempted but failed to displace Austrian rule in much of Eastern Europe, a large number of Hungarians also made plans to relocate to Texas in the years just before the Civil War. Many were soldiers and civil leaders; some were highly educated members of the nobility. Perhaps the best known Hungarians in Galveston were the Vidor family, descended from Charles Vidor who arrived in the city from Hungary in the 1850s. In a story more typical of western movies than real life, Vidor was the only Hungarian in Galveston in 1855. He worked as a clerk for one of the city's most successful merchants and landowners. Charles Vidor married his boss's daughter Emily in October 1858, but sadly, two of their children died in infancy, and the young wife herself died shortly thereafter (McGuire 1993, 136). Following these multiple tragedies, Vidor fought in the Civil War, then invested in the railroad business. By a later marriage, he had a son who succeeded in the lumber business, and a grandson who became a Hollywood legend. But most Hungarians who arrived at the port of Galveston didn't stay for three generations, as did the Vidor family.

Despite the arrival of these new immigrants from Eastern Europe, according to a hand count from the 1880 manuscript census, Germans remained the largest immigrant group in Galveston and in Texas. Other groups of foreign-born immigrants—Italians, Greeks, Belgians, Danes, Mexicans, Portuguese, Spanish, Swedish, Welsh, Canadians, Swiss, Scots, Irish, English, and French—were also counted from the list of names contained in the 1880 manuscript census. Residential locations of some of these foreign-born immigrants are shown in maps 4.3 through 4.7.

AFRICAN AMERICAN SETTLEMENT IN GALVESTON AFTER THE CIVIL WAR

Between 1862 and 1865 the editor of the *Galveston Daily News* discussed issues related to the Emancipation Proclamation with some frequency. Therefore, African Americans and others in the city of Galveston must have known about their impending freedom before the official announcement at the port on June 19, 1865. By then a major increase had occurred in the African American population of the city of Galveston. African Americans made up only 16 percent of the population in 1860 but had increased to 22 percent in 1870. Despite this increase in numbers, the overall segregation of this racial group continued to intensify well into the 1880s. According to recent research conducted by Alwyn Barr, wards in Galveston ranged from 9 percent to 29 percent African American (1995, 66). By 1890 the population of African Americans in Galveston's urban wards ranged from 9 percent to 47 percent (Barr 1995, 104).

Despite the rather large percentage of African American residents in certain parts of the city of Galveston, most continued to reside in back alley dwellings behind the homes of white residents (see Beasley 1996, 22–23). Before alley houses were occupied as primary dwellings, in the first years after Emancipation newly freed African American residents of the city were forced to live in any shack or hut or stable they could find that had survived the Civil War.

> It took little time for alley houses—those structures that were oriented towards the alleys and functionally independent of a front house—to become a primary housing source, especially for blacks. In 1866, Dr. Greensville Dowell, editor of the *Galveston Medical Journal*, wrote, "The negroes are generally living in crowded huts, upon the alleys or in the outskirts of the city." Some of the "crowded huts," converted stables, and alley houses were structures that had survived the Civil War. Others, perhaps most, were built in response to the postwar housing demand. (Beasley 1996, 23)

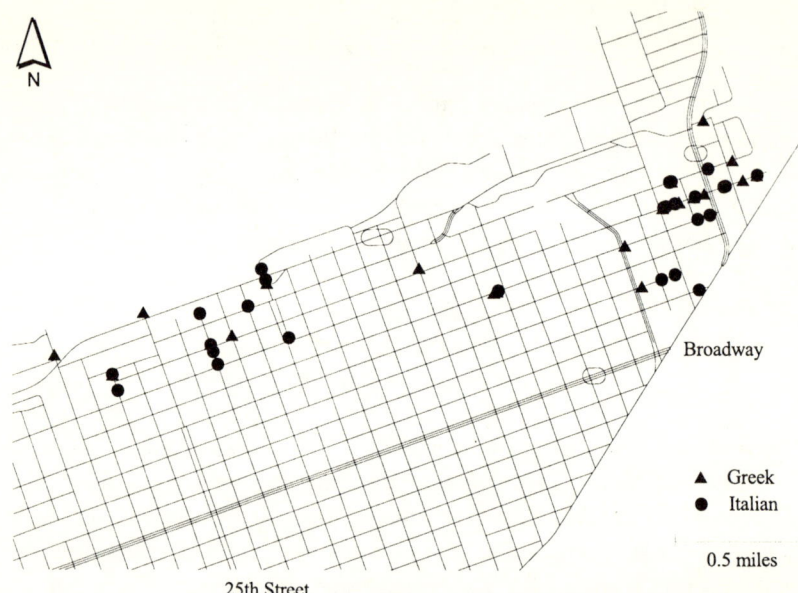

MAP 4.3. *Greek and Italian residential patterns, 1880. (Cartography by Linda F. Prosperie)*

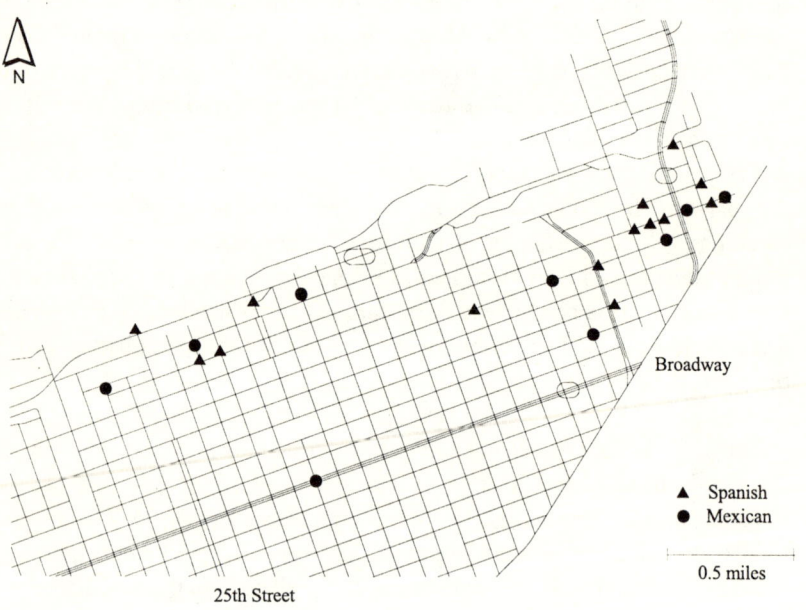

MAP 4.4. *Spanish and Mexican residential patterns, 1880. (Cartography by Linda F. Prosperie)*

RISE OF THE ELITE

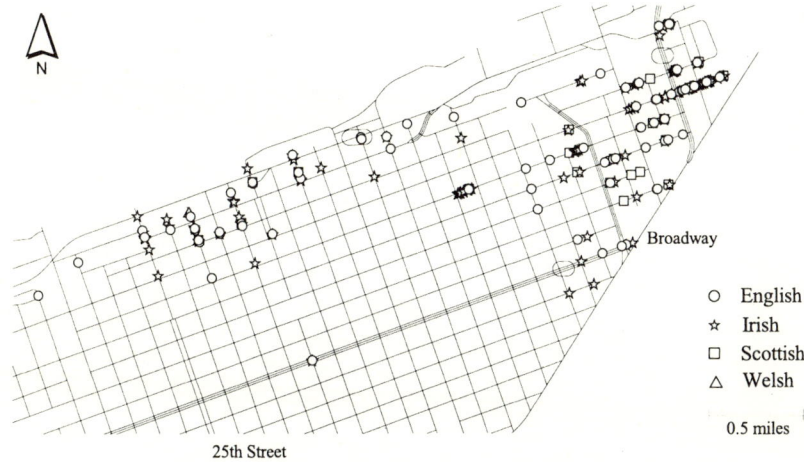

MAP 4.5. *Residential patterns of immigrants from Great Britain, 1880. (Cartography by Linda F. Prosperie)*

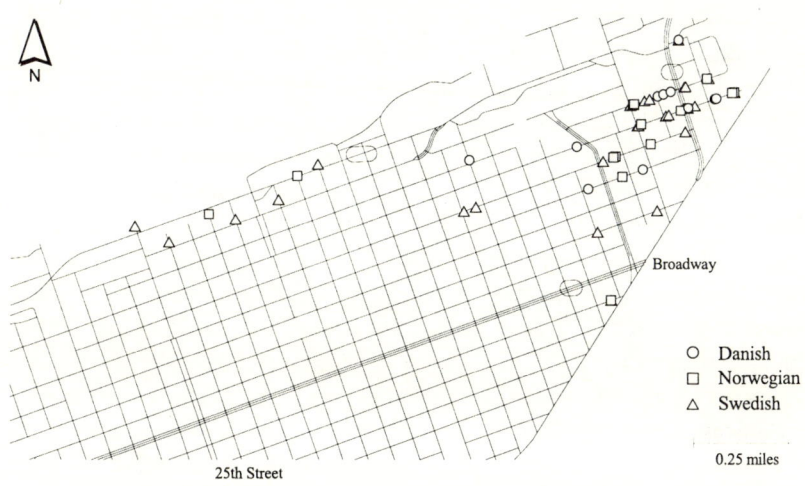

MAP 4.6. *Scandinavian residential patterns, 1880. (Cartography by Linda F. Prosperie)*

Economic and social segregation exacerbated this residential segregation, as separate social and fraternal groups were established in the African American community. In keeping with this pattern of social segregation, the Negro Masonic Lodge, the first organization for African Americans in Galveston, was established in 1875.

MAP 4.7. *Immigrant residential patterns, 1880 (all groups). (Cartography by Linda F. Prosperie)*

Separate religious and educational institutions emerged as well, especially Baptist and Methodist. The Freedmen's Bureau established the first "colored school" in Galveston in 1865. Most of the teachers for the school were provided by the American Missionary Society. In an effort to attract teachers to move to Galveston, this text was distributed as part of a recruiting pamphlet: "appropriate to women's special faculty and adaptations . . . here is a demand for her sweetest sympathies and her boundless charity . . . how much better this than to be groping about in the dark, as so many of them are, seeking to answer the craving for work in callings that belong more properly to men" (*Journal of the American Missionary Association* [May 1867]: 104).

By 1867 eight teachers, most from the New England states, had arrived in Galveston to teach at the "colored" school. Most found boarding with German families even though "their homes received threats of having their houses burned . . . and . . . the majority of the Southern white non-German community ostracized the teachers and considered it a sin and disgrace for a member of their own race to teach blacks" (*Journal of the American Missionary Association* [May 1867]: 104). By 1867 the Freedmen's Bureau had established a high school, four day schools, and three night schools, as well as a normal school to train teachers. Both children and adults attended classes at these all-African American schools. By 1869–70 the Texas Constitution provided for free public schools and made attendance compulsory for all children from the age of five to eighteen years, regardless of race. The scholastic census in Rosenberg Library in Galveston records 2,478 white students and 631 African American students enrolled in Galveston public schools in 1870, the final year the Freedman's Bureau operated public schools on the island. Thereafter most attended "black only" public schools, until Catholic bishops and Dominican nuns opened a school for African American children in the city in 1886.

A moving symbol of the evolution of educational opportunities for African Americans who lived in Galveston is Central High School. This school for African American students opened as a storefront in 1885. Although the school's curriculum has not been documented, the aim was to further the education of its African American students in sixth and seventh grades. By the following year, eight of the students had passed examinations that qualified them for admission into high school; the first African American high school in Texas, Central Colored School of Galveston, was opened at the corner of Avenue N and Fifteenth Street to serve their needs (Jackson 1998b, A1).

Central High School was the all-important link between elementary school and higher education for African American students in Galveston, and thus a major symbol of the city's support for black families. As it grew, a new brick building in the Gothic style was built at the corner of Twenty-sixth Street and Avenue L. In 1904 the school was selected by the city to house a branch of the Rosenberg Library. After the school's accreditation in 1926, graduates were able to apply to college without taking eligibility exams. This decision opened the door for the establishment of a series of African American colleges in Texas.

A new building in a new location was constructed at Thirtieth and Sealy for Central High School in 1954 in an effort to make the facilities and curriculum in the African American school equal to those of the white Ball High School in Galveston (table 4.1). Ten years later, when education in the city was desegregated, Central High School became a junior high school, which opened its doors for the first time to a mixed group of students. This same building today is Central Middle School and has an ethnically and racially diverse population of students. The original building at Avenue L and Twenty-sixth Street is now the site of the Old Central Cultural Center (fig. 4.5), presided over by one of its original graduates.

Churches were also segregated. As early as 1848 the Methodist Church in Galveston had bought a lot on Broadway between Twentieth and Twenty-first Streets for slaves to have outdoor church meetings. The Methodists erected a building at this site in 1863, the first of its kind in Galveston. Ownership of the building and lot was transferred in 1865 to African Ameri-

TABLE 4.1. A Micro-View of Social and Political Change in Galveston: A Chronology of Central High School

1885	Central High School was established on September 15, 1898, occupying three floors of a storefront on the southeast corner of Sixteenth Street and Avenue L. C. John Waring was principal.
1889	John R. Gibson became principal, and Central Colored School of Galveston was housed in another wooden storefront-type structure on the corner of Fifteenth Street and Avenue N.
1893	A brick building was erected to house the high school at the corner of Twenty-sixth Street and Avenue M.
1904	Central High School was selected as the site for a branch of Rosenberg Library.
1905	The Colored Branch of the Rosenberg Library opened.
1924	Extensive improvements were made, and a new wing with additional classrooms and labs was added to the school.
1926	Central High School became accredited by the state of Texas. This allowed graduates to enter college in the state without having to take an exam to prove eligibility.
1933	Central High School became affiliated with the Southern Association of Secondary Schools and Colleges.
1936	Principal Gibson retired and was replaced by Walter J. Mason.
1941	Leon A. Morgan was appointed principal. Morgan remained principal until the school was desegregated in 1968.
1954	Central High School moved to a new campus that stretched from Thirtieth to Thirty-third streets between Avenues H and I.
1965	Galveston School District desegregated all of its facilities, bus transportation, and student activities.
1968	Central High ceased to exist when its students were merged with Ball High students to achieve complete integration.
1976	The Central High School building at Twenty-sixth Street and Avenue M was renovated as the Old Central Cultural Center as a part of the American Bicentennial Celebration.
1977	A historical marker was placed on the property.

Source: Jackson 1998b, A6.

can members of the congregation, who then organized the African Methodist Episcopal Church (Scheibe 1992, 92). It was here that a crowd of between eight hundred and a thousand African American residents of Galveston celebrated their emancipation on January 1, 1866, when "the pastor read the Emancipation Proclamation and General Edgar Gregory, director of the Freedmen's Bureau spoke to them. The day was s stormy one, rain filling the streets with mud and water. The crowd sang 'John Brown's Body' before the celebration ended" (*Galveston Weekly News*, 10 January 1866, 1; *Flake's Bulletin*, 2 January 1866, 1).

This church was later named Reedy Chapel A.M.E. Church, after the first pastor of the church. This still present spiritual and social center remains an important part of the African American cultural landscape in Galveston (fig. 4.6).

Between the end of the war in 1865 and the yellow fever epidemic of 1867, the Freedmen's Bureau increased school attendance of African Americans in Galveston 400 percent by abolishing tuition. Curriculum in African American schools, as in other schools in the city, consisted of geography, reading, writing, arithmetic, and needlework, with Bible stories used for grammar lessons.

Despite these efforts to educate and support newly freed slaves in the city, African Americans in other parts of southeast Texas didn't fare as well during the Reconstruction years. A vivid example of the difficulties of surviving

FIGURE 4.5. *Central High School at Twenty-sixth Street and Avenue L provided educational opportunities for Galveston's African American students. (Courtesy of the Rosenberg Library, Galveston, Texas)*

FIGURE 4.6. *Almost a thousand African Americans celebrated the announcement of their Emancipation at this African Methodist Episcopal Church in Galveston on January 1, 1866. (Courtesy of the Rosenberg Library, Galveston, Texas)*

in the postwar South is the effort to leave the country organized by African Americans in east Texas in the 1880s. This group appealed for congressional appropriations to finance their emigration to Liberia, but their request was denied. Another group of African Americans living in San Antonio tried unsuccessfully to organize a steamship company that would sail out of Galveston. More than ten thousand African Americans did resettle in Mexico during the late 1880s, before the Mexican government canceled their contract. Again in 1899 rumors of ship departures to Liberia from Galveston circulated in Texas, bringing many hopeful emigrants to the port of Galveston, only to be discouraged once again (Barr 1995, 97).

SOCIAL AND ECONOMIC IMPACTS OF ETHNIC AND RACIAL DIVERSITY

The hard work and frustration of key members of Galveston's laboring class, many of whom were new immigrants and African Americans, led to the city becoming the earliest location of union activity in the state. As early as the 1850s the Galveston Typographical Union No. 27 had been organized, followed by the better known Carpenters' Local No. 7 in 1860 and the Screwmen's Benevolent Association in 1866. Galveston had the only two legitimate unions in the state during the Civil War. In 1895 the Galveston city directory listed over thirty labor-related organizations in the city.

The successful effort to organize unions in a place like Galveston is surprising considering the city's dependence on immigrant and African American labor and its long-term dominance by a powerful white, elite business class. The thirty-four original members of the Screwmen's Benevolent Association were a highly diverse group. Only five were native-born U.S. citizens (two were from New York, two from Philadelphia, and one from Boston). Of the foreign-born members, eleven were Irish, ten were English, six were from Scandinavian countries, and one each came from Scotland and Germany (Reese 1971, 161). Although all were able to sign their names to the union roll book, none was among the more financially secure residents of the town. Fewer than five maintained a residence; most lived in hotels and rooming houses. None played any role in civic or cultural affairs in Galveston. But these first members were typical of the membership of the Galveston Screwmen's Union over the years, and thus representative of the working-class residents of the city.

Screwmen were skilled longshoremen who used a jack-screw, a ratchet tool weighing 250 pounds that could be braced against a bulkhead to compress cotton bales on board ship and increase the cargo capacity—thereby also increasing the profit to the ship owner. This special skill and role at the wharves gave the screwmen a competitive edge in the hiring pool over other laborers, at least until cotton compresses were developed in the early 1900s.

The membership of the Screwmen's Union was always dominated by foreign-born workers. According to Reese, there was a shift in the national origin of union members, from the first members in 1866 up to 1900, from Irish and English to German and Scandinavian—even though there was never "more than a handful of members who were born in eastern or southern European countries" (1971, 162).

The unions in Galveston were active organizations, with a predilection for strikes founded in dissatisfaction with wages. Because salaries varied in different places and in various occupations, strikes soliciting wage hikes were common. They were also quite often unsuccessful.

The unions played a significant role in bringing racial inequality to the attention of island residents. African American laborers were quite often used as scabs to replace white laborers when they were out on strike. Conversely, African Americans were also seen as the persecuted class in labor disputes. Two years after a potentially violent strike of all African American workers in the city was averted with the help of local politician and moderate Norris Wright Cuney, in 1879 a group of African American longshoremen established a Cotton Jammer's Association. This was the first all-African American union in Galveston. In 1883 Cuney and this organization managed to finally break the monopoly held by white dockworkers by securing a contract that allowed African American workers to be employed at the large Morgan Wharf. Having gained a foothold, Cuney and his African American workmen then moved against the Screwmen's Union, organizing an all-African American union called the Screwmen's Benevolent Association No. 2 (Reese 1971, 180).

In 1882 a major strike by 280 cotton handlers in Galveston failed to prevent the hiring of African American laborers. The following year, white longshoremen organized a strike against the Mallory Line to protest unacceptable working conditions and also Mallory's hiring of nonunion African American workers. Racial tension on the Galveston docks continued for many years following this strike and a second one in 1884. Both times, unions protested management's use of African American scab labor and low wages. African American workers, however, benefited from the outcome of these strikes because they were ultimately able to gain some control of the labor situation on both the Mallory and Morgan Wharves. But a wage strike by African American union men in 1898 resulted in the loss of many of these jobs—rather than bringing an increase in wages as they had hoped (Kelly 1975, 84–85).

In 1906 Galveston was selected over New Orleans as the site of a major new immigration station. The planned location was Galveston's Pelican Island—copying the successes of Ellis Island in New York. This building was never built, although a smaller quarantine station was constructed there in the 1920s (fig. 4.4). In 1913 a scaled-down version was constructed on the Strand. It was damaged by high winds in 1915 and closed the following year. The building occupying this site today was built in 1933. It originally housed a dorm, medical facility, kitchen, and dining and recreation room for new immigrants. In 1940 this building was converted to a U.S. customs house, and it is now owned by the University of Texas Medical Center.

CONCLUSION

As the era of Reconstruction came to an end, and railroad lines, improved plank roads, and steamship connections developed in postwar Texas, Galveston grew and prospered. Due to these vastly improved transportation linkages, the complete rebuilding of the city after the war, and expansion of its residential, commercial, and industrial sectors, the city continued to serve as the state's most important port (and the Gulf Coast's second busiest port, exceeded in tonnage only by New Orleans).

City officials decided to take action to stem the tide of Houston's growth at Galveston's expense. In the mid-1890s a resolution was passed to improve Galveston's port. The resolution had the support of inland cities, such as Denver and Fort Worth, which were connected to Galveston by railroad lines and thus dependent on it for importing and exporting maritime products; the measure even won the support of the Houston City Council. In 1896 a 25-foot-deep channel was dredged that cut through shallow sandbars and thus deepened the harbor (McComb 1983, 17). Many believed that such major improvements to the port of Galveston would solve all its problems and return it to a position of power and dominance in the state and in the Southwest.

Despite this attempt to create an improved port facility, which once again made Galveston the major distribution center for the interior for a brief period, other problems lingered. Most centered on the dramatic transformation of the city's economic landscape. Although the city was essentially the same size and had the same morphology in the postwar years as it had earlier, albeit larger and more prosperous, a major and almost invisible change had occurred in its economic structure. This economic shift would affect not only working-class residents of Galveston, but also its future as the dominant city in Texas. One of the most serious factors in this shift was the attitude and conduct of the powerful Galveston Wharf Company, central to development of the port of Galveston through the second half of the nineteenth century.

Describing economic change in the twenty years immediately after the war, Kelly (1975, 76–77) reports that mercantile houses, which had previously carried supplies needed by planters, began to focus their efforts on East Coast and European buyers to provide their shoppers with a wider selection. Galveston thus evolved into an important wholesale market for merchants in interior cities who formerly had had to import their stock from New York or New Orleans.

Galveston's ever expanding combined wholesale and retail trade was estimated at only $18,320,000 in 1870 and had increased to $47,300,000 by 1885 (Morrison 1887, 30, 65, 67). The move to consolidate power in the hands of fewer merchants must have seemed like a distinct step in the right direction to Galveston business owners. Undoubtedly, the island profited greatly in these years by the expansion of Galveston firms as they continued to channel increasing amounts of money into the local economy.

The move toward specialization and business expansion in Galveston had its price, as the growth of monopolies among the city's trading houses increased in the second half of the nineteenth century (Kelly 1975, 78). Examples of this move toward monopolistic business abound. Chief among them was the control of sales of consumer goods, such as hats and caps, boots and shoes, and other types of clothing. Similarly, the *Galveston Daily News* served as a clearinghouse for all news in the state up to 1885, and the port of Galveston had an absolute monopoly on the use of cotton compresses until the mid-1870s (Kelly 1975, 77).

This economic power found its focus in the evolution of an upper-class, white, native-born business elite in the city. The formation of the powerful Galveston Wharf Company is the best example of this consolidation of power among the city's wealthiest families. This company was called the "Octopus of the Gulf" and "hydra-headed monster" by its detractors, because it charged outrageously high fees at the port. The Galveston Wharf Company itself publicly criticized the evolution of monopolistic control on the island—yet it ultimately developed policies that ensured its own total control over all wharf facilities and trade (Sibley 1968, 85, 87, 97).

First organized in 1854 by a small group of prominent Galveston business owners, the Galveston Wharf Company ultimately held complete power over all wharf facilities, ownership of the waterfront area, and monopoly control over all policies relating to Galveston's imports and exports. Under pressure, it finally admitted the city of Galveston as a silent partner in 1869, but this did not relieve the criticism from outside and from the *Galveston Daily News* about the exorbitant port charges (McComb 1983, 15).

The Galveston Wharf Company provides a vivid example of the sharp boundaries in Galveston that divided native-born from foreign-born residents and separated differing socioeconomic groups that were so much a part of everyday life on the island up to the time of the Great Storm in 1900. Galveston's wealthiest and most powerful citizens had total control over an institution that normally would have been a public utility. The Galveston Wharf Company monopolized the Galveston wharves after 1854, controlled the Santa Fe Railroad when it was built in the 1870s, and was closely allied through its investors with the Mallory Line, cotton warehousing, and even the Galveston Cotton Exchange.

From this one example, then, it can be seen that a few families controlled the economic power in the city of Galveston and at its all-important port. Below these few wealthy investors was a mass of white, middle-class citizens—cotton merchants, steamship agents, insurance brokers, and retail merchants—whose place in society was determined by their place in the island's hierarchy. Their labors made the control by a few families possible. According to Knox, "wealth begat power, power begat wealth, and both wealth and power descended from the aristocracy. But there were citizens without power and wealth. Small businessmen, clerks, and laborers had as their duty to follow their leaders. When they did that, the municipality was at peace. When they did not, the aristocracy responded with vigor . . . the aristocracy always remained in control" (1983, 8).

Despite memories of the news of the devastating 1875 hurricane that had severely damaged the port city of Indianola, also located on the Texas Gulf Coast, ever confident Galvestonians believed their city would grow and prosper forever. Even when a second hurricane wiped out Indianola in 1886, and despite repeated warnings publicized by the *Galveston Daily News*, most residents of Galveston believed their city would be spared. They rested their firm belief in the theories of a navy oceanographer who had once promoted his ideas that Galveston Island did not lie in the path of killer storms and so would be protected from serious threat.

Economic growth and optimism about the future was indicated by figures published by the U.S. census in 1890. That year Galveston's total population was 29,084, making Galveston the state's third-largest city, after Dallas and San Antonio. Despite its fall from the number-one position that year, when the new University of Texas Medical Department in Galveston started classes in the fall of 1891, Galveston had never been more prosperous. Its commercial and residential landscape continued to become more densely developed and more refined throughout the 1890s as a few of the city's earliest residents donated money for ongoing beautification of the city.

Other problems, more pressing than dealing with an aristocracy, would prove to be even more disastrous for Galveston Island than poor planning by power-hungry corporations. The major hurricane that struck the city in September 1900, discussed in chapter 5, was only one in a long list of issues affecting the ongoing economic development of the city. Horrendous damage to property and the death of more than six thousand Galveston residents caused by the Great Storm have already been recorded in numerous other sources, most recently in Larson's book *Isaac's Storm*, published in 1999. Nowhere to date, however, do scholars or other observers of post-1900 Galveston comment on the impact of the storm on the city's residential and commercial patterns. Great change was on the horizon for the city's lower-income residents in the years following the storm. Not only would the island's whole approach to government change, so too would its physical and human landscapes.

5
After the Storm
REINVENTING AN ISLAND ENVIRONMENT

The accounts of [the Great Storm] are given to me by my grandmother . . . about my great-grandparents, Joseph and Margaret (Maggie) Devoti. . . . Joe was 24 years old and Maggie was 20 years old at the time of the storm. Joe worked many jobs, and in 1900 he worked at a plum orchard around what is now 59th and Heards Lane, which helped to pay rent on a house they lived in around that area. They had three children at the time—Rebecca, 4, Hazel, 2, and Louesin Pearl, three months.

There was no Seawall and the Gulf and the Bay met. When the water started to come in the house, they did everything they could to hold on to the children, but the raging water swept the children from their arms. Joe and Maggie got separated.

Joe managed to climb up in a cedar tree around 59th and Heards Lane. He saw a woman . . . he didn't know if she was dead or alive in the water. She was floating on a table with her long red hair wrapped around one of the legs. He managed to pull her up by her hair into the tree with him, and to his surprise it was his wife, alive. There were bodies and animals in the water everywhere. Maggie was most afraid of being eaten by a pig.

They stayed in the cedar tree for three days with no clothes, no food, and nothing to drink. The Coast Guard picked them up after the third day and gave Maggie a red soda to drink. She said it was the most delicious thing she had ever tasted. Maggie wanted to leave the island for a while, but Joe said he had to look for the girls. He found one of the little girls dead, and identified her by her shoes being put on wrong by him in the rush before the storm. He put her little body in a baby buggy he had found, and took her to the Old Catholic Cemetery on Broadway to be buried in the Devoti section.

Joe's sister Catherine Devoti was married to Paul DeLaya who was the Chief of Detectives in Galveston at the time of the storm when he perished. Catherine and their three sons survived and left to go live in San Antonio, Texas.

Joe's other sister, Louise Devoti, was married to Thomas Millo. They had

four children. Before the storm hit, Louise put her daughter Lucille and her son Charlie up in a tree. She could not climb the tree and perished with her other two children. Thomas Millo was downtown at the time of the storm and survived. We lost twenty-five blood relatives in the storm.
MARGARET DEVOTI COOK, "THE 1900 STORM,"
STORM FORUM, WEB SITE OF THE *GALVESTON DAILY NEWS*, 2000

Warmest climes but nurse the cruelest fangs . . . the Typhoons will sometimes burst from out of that cloudless sky, like an exploding bomb upon a dazed and sleepy town.
HERMAN MELVILLE, *MOBY DICK*

It must have felt like the cruelest "typhoon" of them all on that Saturday afternoon in September 1900. The unexpected Great Storm of Galveston, the greatest natural disaster in North American history, left its mark upon the culture and the shape of the city of Galveston. By the turn of the twentieth century, the city had evolved into a diverse urban place with residents from many parts of the world. Table 5.1 makes it clear that the island had become increasingly cosmopolitan since 1880. One of the immigrant groups that suffered the greatest losses in the aftermath of the devastating hurricane was Galveston's Italian community.

THE GREAT STORM AND ITS SURVIVORS

In the decade before the storm, Galveston had reached its zenith of prosperity. Although the census of 1890 revealed that both Dallas and San Antonio had for the first time surpassed Galveston in size, the island city was still the state's most important trading and cultural center. Galveston's mixture of ethnic and racial groups and its diverse cultural landscape made it the most culturally rich urban center in the state. Throughout the 1890s cultural activities, such as regular performances held at the ornate Grand Opera House, were a regular part of island life. Adding to its reputation as the state's elite city was the opening of the University of Texas Medical Department in 1891.

The city was booming economically as well. Galveston was the number-one cotton port in the nation at the end of the 1890s. Only four other ports were handling more tonnage overall. Indeed, Galveston was at the peak of its prosperity and influence in the years just before the storm (Miller 1983, 140). It was a global city, maintaining international connections through its

MYTHIC GALVESTON

TABLE 5.1. **Foreign-born Population of Galveston, 1900**

Asia	5	Mexico	152
Austria[a]	281	Norway	189
Belgium	18	Poland (Austrian)	37
Bohemia	18	Poland (German)	9
Canada (English)	156	Poland (unknown)	36
Canada (French)	29	Russia	3
China	67	Scotland	164
Denmark	92	Spain	138
England	873	Sweden	60
France	357	Switzerland	394
Germany	2,450	Turkey	83
Holland	27	Wales	36
Hungary	49	West Indies	57
Ireland	834	Others	165
Italy	560	At sea	11
		Total foreign-born	7,328

Source: U.S. Census of Population, 1900.

[a] Note, according to Jordan 2000, that "Austrians" were, in fact, Czechs, but that a more accurate Czech count would be all Bohemians plus Austrians.

active import and export trade and also through its busy immigration processing center on Pelican Island. According to the Galveston city directory published in 1899–1900, the year before the storm, there were foreign consulates in Galveston representing fifteen different nations: Belgium, Costa Rica, Denmark, Germany, Great Britain, Italy, Japan, Mexico, Netherlands, Norway, Russia, Spain, Sweden, Switzerland, and Venezuela. This same directory listed the many different social clubs and societies in the city. Nine of these were organized by Jewish residents, seven by Germans, and one each by British, Italian, and Irish residents of the city. By the time the census of 1900 was taken in the spring before the storm, Galveston had a total population of 37,788, up 30 percent from the previous census count.

In contrast, Houston, only 50 miles to the north, remained a "brawling, hard-luck town located between unpromising prairie to the north and swampy bayous everywhere else" (Mason 1972, 19). In 1900 Galveston residents were quite condescending toward their neighbor to the north. Despite Galveston's railroad connections to the interior and efforts to deepen their inland port on Buffalo Bayou, most island residents believed Houston's struggle to become a great commercial center was doomed to failure.

AFTER THE STORM

Alexander E. Sweet wrote this description of the port of Houston on his first visit: "The Houston seaport is of a very inconvenient size—not quite narrow enough to jump over, and a little too deep to wade through without taking off your shoes. When it rains, the seaport rises up twenty or thirty feet and the people living on the beach, as it were, swear their immortal souls away on account of the harbor facilities. The Houston seaport was so low when I saw it last that there was some talk of selling the bridges to buy water to put in it" (Mason 1972, 20–21).

PERCEPTIONS OF THE ISLAND'S POTENTIAL FOR DANGEROUS HURRICANES

With a centuries-long Gulf Coast record of destructive and highly dangerous storms, it is difficult to imagine why anyone would have chosen this small, flat island as an appropriate settlement site. Hurricanes and other major storms had been recorded for over a century before the Great Storm.

Historic Storm Events

Hurricanes that reach the Texas Gulf Coast form anywhere from the Cape Verde Islands off the coast of Africa to 150 miles from the Texas coast. Between 1810 and 1886 sixteen major storms damaged Galveston Island (table 5.2). Although the Great Storm holds a unique place among the city's historic storms, there were many previous storm events indicating the serious risk associated with settlement of the island.

Storms usually approach the Texas coast at right angles, although they have been known to travel parallel to the coast. This is an important factor in explaining why so much damage potential exists on the island. The direction of the storm's approach influences the height of the storm surge, one of the leading causes of hurricane damage. Storms approaching at right angles increase the height of the storm surge because of the additive effect when they coincide with regularly occurring tides. In addition, flood waters resulting from storm surge generally rise faster than normal high tides. And areas undergoing land subsidence—such as the Houston-Galveston region—are at great risk from flooding associated with these storm surges. Paine and Morton (1986) compiled a table of maximum surge heights from hurricanes near Galveston Bay and found that heights reached 20 feet in the storm of 1900 and approached or exceeded 10 feet on eight other occasions up to 1983 (table 5.3).

TABLE 5.2. **Tropical Storms in Galveston 1810–86**

Year	Remarks
1810	Reported by local Indians, "water as high as four Men"
1815	Reported by local Indians
1818	Ruined Lafitte's camp, fort, and residence
1829	Immigrants wiped out by storm
1837	"Racer's Storm," nearly everything in town destroyed
1839	Lasted three days and damaged shipping
1842	Considerable damage to buildings and ships
1854	City flooded, considerable damage from hurricane
1866	Hurricane
1867	About $500,000 worth of damage and a "number of lives lost"
1871	Considerable property damage
1871	Extensive flooding and severe losses in shipping; some loss of life
1874	Little or no damage
1875	Tropical storm. About 20 killed, $200,000 property damage, storm surge made 3 huge cuts into island
1886	Damage along beach; ship and crew lost
1886	Tropical storm. Six lives lost; 12–18 inches of water in business section; $150,000 damage

Source: Adapted from Claflin 1985.

Since 1983 several major storms have damaged the Gulf Coastal region and its offshore islands. In 1983 Hurricane Alicia struck Galveston's west end, forcing the beach boundary back at least 150 feet. In October 1989 a "Category 1" storm named Jerry led to the death of three people on Galveston Island, who drowned when their car was driven off or blew off the seawall in a blinding rainstorm. In September 1998 two people on Galveston Island died as a result of the winds and high waves created by Tropical Storm Frances: a father and daughter who were swimming in the surf during the second day of the storm. Frances caused the most severe flooding in Galveston since Alicia in 1983.

In light of today's documentation of these historic storms and predictive scientific information about the potential for frequent and regular storms affecting Galveston's land and people, it is easy to blame early developers for ignoring the potential damage to the island. Like so many people who live in dangerous environments today, however, most Galveston residents in those days viewed the risk from hurricanes as nonthreatening and exagger-

TABLE 5.3. Hurricane Surge Heights (feet), 1837–1985

Date	Surge Height	Location
1837	5–7	Galveston
1847	8–10	Galveston Channel
1854	9–10	Galveston Channel
1867	6.6	Galveston Channel
1875	8.3	Galveston
1886	9	Galveston
1886	6–7	Galveston
1900	20	Galveston
1909	10	Galveston
1915	14.3	Galveston
1915	16.1	Galveston Causeway
1919	8.8	Galveston
1933	6	Galveston
1942	6.2	Galveston
1949	7	Galveston
1957	6.2	Galveston
1961	9.3	Galveston
1961	15	Upper Galveston Bay
1983	7.4	Galveston Channel

Source: Adapted from Paine and Morton 1986.

ated. From the earliest days of settlement, newcomers to the island heard—but usually ignored—stories about storms that had caused problems in the past. According to Jordan, "even the earliest settlers had heard of the storm in 1818 which sent ocean waters surging completely over Galveston Island, and the force of these hurricanes was obvious from wrecks of old vessels which were found in timbered country five miles from the seashore in the 1820's" (1980, 14).

In the days just before the Great Storm, storm warnings were not specific enough to adequately warn Galvestonians of the impending danger. After the storm hit Cuba, it was thought to be heading out into the Atlantic. Some climatologists thought it was headed for Louisiana. On Galveston Island storm flags were hoisted to warn residents, but the intensity of the storm was unexpected. Those who had survived the lesser storm of 1886 were unconcerned. With no Doppler radar, no satellite system, and no airplanes to measure the storm's direction and intensity, there was no means to make an accurate prediction of the "direct hit" on Galveston.

IMPACT OF THE GREAT STORM ON PEOPLE AND PLACE

The first book published locally in Galveston after the hurricane contained this disturbing description:

> A city of stunned and stricken people, a thousand naked, five thousand bruised, ten thousand homeless, all searching for their friends among the slain, tearless but bleeding at the heart, unappalled and uncomplaining; a city of wrecked homes and streets choked with debris sandwiched with six thousand corpses; a city leafless and bloomless, with the slime of the ocean on every spot and in every house; a city with only three churches standing, not a school building or benevolent institution habitable; a port with shipping stranded and wrecked many miles from the moorings, light-craft cast upon the land as the playthings of the elements, ocean-going vessels sent crashing through railroad bridges or driven high onto shore and shallow. (Ousley 1900, 23)

As early as five o'clock that Saturday afternoon, the tide was about 9 feet higher than average and covered the highest streets. At the height of the storm, waves were at least 15 feet high at the beach. The velocity of the sustained wind was recorded at 84 miles per hour on Saturday evening, just before the anemometer blew off the top of the Levy Building at Weather Bureau headquarters. Before this damage to instruments occurred, gusts up to 100 miles per hour had been recorded in the heart of the city's central business district.

Strangely enough, warnings about the storm's potential danger had been predicted by the U.S. Weather Service in Washington, D.C., each day for more than a week before its arrival at Galveston. One might wonder why residents of the island didn't leave immediately when they heard about the forecasts that were communicated in the daily paper several days before the storm. But the last railroad off the island left at noon. By dark it was too dangerous to leave. Evacuation routes were problematic or impossible because of the insular location of the city of Galveston and its limited connections to the mainland.

No doubt many people were unafraid and misjudged the protection their homes could provide for them and their families. Other, much less powerful storms had blown through their lives before—and they had survived them all. Still others believed the gentle slope of the island's Gulf-front beaches would dissipate the force of the waves before they reached shore.

The storm blew through the city relatively quickly. The head of the local weather bureau, Isaac Cline, estimated that winds reached at least 125 mph later that evening. According to Larson's account, Cline opened his front door at 6:00 P.M. and saw "a fantastic landscape. Where once there had been streets neatly lined with houses there was open sea, punctuated here and there by telegraph poles, second stories, and rooftops . . . the sea was strangely flat, its surrounding face blown smooth by the wind. By 11:30 P.M. that night, the wind quieted and water began to recede. But the calm after the storm was almost worse than the terror of the hurricane itself" (1999, 188).

The next morning, Sunday, September 9, dawned clear and calm. An eyewitness described "Galveston on September 9th, 1900, the sad and fateful Sunday when she awakened to her sorrow, when there was no preaching or public service, but when every soul had faced its God and knew its own accounting, when the first sob and last sob of an unspeakable grief welled once to the lips and then was smothered in the tremendous responsibility of caring for the living, when the meridian sun had scarce begun, and when 30,000 strong men and brave women took up a task which mankind had never been called upon to perform and settled down to their work" (Ousley 1900, 23).

As the stunned survivors crawled out of the wreckage that Sunday morning, they witnessed entire blocks of the city completely destroyed. "The dead were everywhere, thousands of them. Thousands more walked around scantily clothed or even nude and in shock" (Macdonald n.d., n.p.). City streets that had been lined with homes and prosperous businesses one day earlier were cleared except for a few sticks of wood and piles of bricks and rocks. According to Macdonald, "a great wall of debris that in some places measured two stories tall stretched across the city. Comprised of pieces of homes, furniture, cats, dogs, and horses, the wall also contained people, both barely living and the dead. The wall began in the east, wrapping itself around St. Mary's Hospital and then zigzagging southwest toward Sacred Heart Church on 15th and Broadway" (n.d., n.p.). The storm had swept away more than twenty-six hundred homes and reduced a thousand more to piles of rubble. The zone of extreme flooding by Gulf waters during the storm is shown in map 5.1. According to Lowry, damage was not limited to the island but also extended out to sea: "The warehouses along the bayfront were a tangled mass of broken timbers, and the area was littered with overturned and beached vessels. The British ship *Taunton* was stranded 22 miles away in Trinity Bay and the steamer *Roma* was torn from its moorings, pushed into the steamship *Kendal Castle*, and then blown sideways down the channel, taking out three bridges along the way" (2000, 48).

MYTHIC GALVESTON

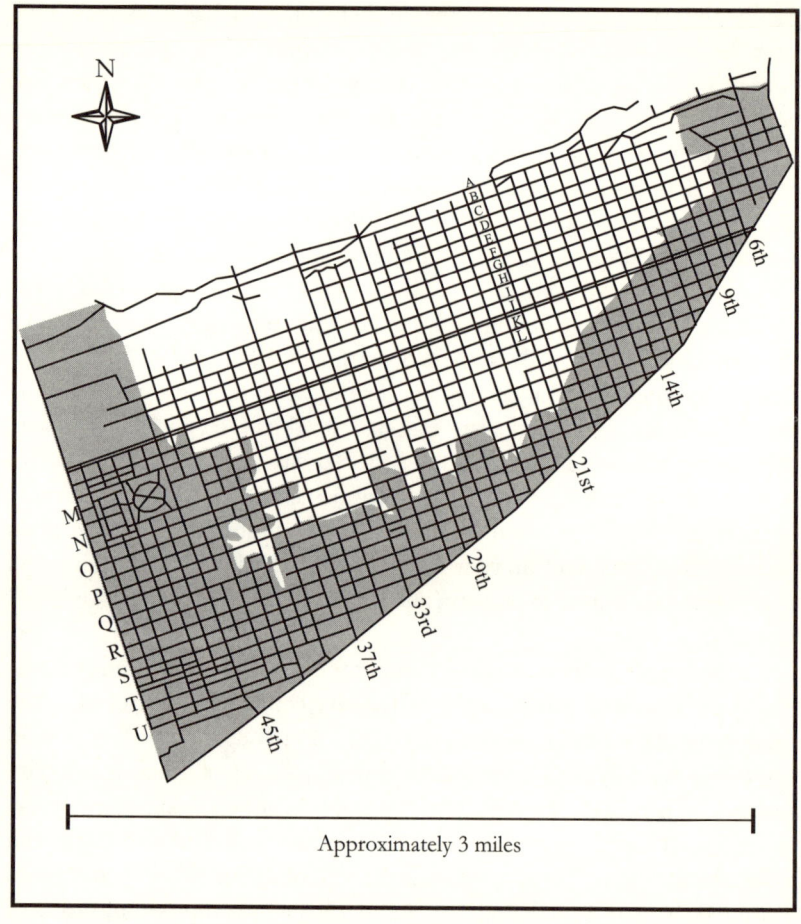

MAP 5.1. *Area of total devastation (shaded) after the Great Storm, 1900.*
(Cartography by Linda F. Prosperie)

As map 5.1 shows, in 1900 Galveston was developed up to its beach line. The residents who lived south of Broadway toward the Gulf of Mexico lived in small houses that absorbed the onslaught of the hurricane. All their homes were lost. If they did not evacuate into the interior of the island, all of these residents died in the storm. Italian and other new immigrants lived in this part of the city in large numbers (see map 5.4).

Thirteen-year-old Will Murney was one of the ninety-three children trapped in St. Mary's Orphanage, located on the edge of the Galveston beach, on the night of the storm. His eight-year-old brother Joseph was also living at the orphanage. Will survived; Joe did not. Two other orphans were saved—Frank Madera and Albert Campbell. All the other orphans died in the storm. In a letter written a month after the storm, young Albert said, "I do remember being on a roof or something. I saw boys and girls on a roof. I remember also one of the men at the home having a baby on his back. But it washed off. I never did remember getting on a tree . . . I do remember getting down off the tree . . . warned teacher to keep it from children regarding the flood" (Macdonald n.d., n.p.).

After the Great Storm, nothing could ever be the same again. A few days afterward, the local paper reported that the first casualty had been an African American baby who had fallen out of the window of a house on Twenty-first Street near Avenue N. According to this terse report, "The wind and waves had backed the water up in the ditches and into the front yard, and it was sufficient in depth to drown the child" (*Galveston Daily News*, 13 September 1900, 2).

THE EFFECTS OF THE STORM ON CITY INSTITUTIONS

Damage to schools, churches, and orphanages in Galveston from the Great Storm was severe. This was particularly disturbing to survivors because religious and educational institutions tend to hold communities together, especially in times of stress and duress. These three all-important types of institution are shown in map 5.2 as they appeared before the storm and in map 5.3 as they appeared two years after the hurricane. Accounts tabulated immediately after the storm documented that twenty-two Protestant churches in Galveston were totally destroyed and the remaining twelve severely damaged. Of the Catholic churches in the city, St. Patrick's on the city's west side was completely destroyed. According to Father James Kirwin, writing about the beauty of this cathedral, which served so many of the Irish immigrants and other Catholics in the city, "No brush can ever paint nor pen describe that 'scene so sad and fair'" (Ousley 1900, 101).

Likewise, the Jesuit Sacred Heart Cathedral, which served as a haven for more than four hundred refugees during the storm, was severely damaged. This large building—even in a damaged state—along with other Catholic institutions on the island, such as the Ursuline Convent (where four babies were born the night of the storm) and St. Mary's convent, provided shelter for survivors of the storm. At St. Mary's Orphanage, located on the beach at Sixty-ninth Street and Central City Boulevard, all but one of the ten sis-

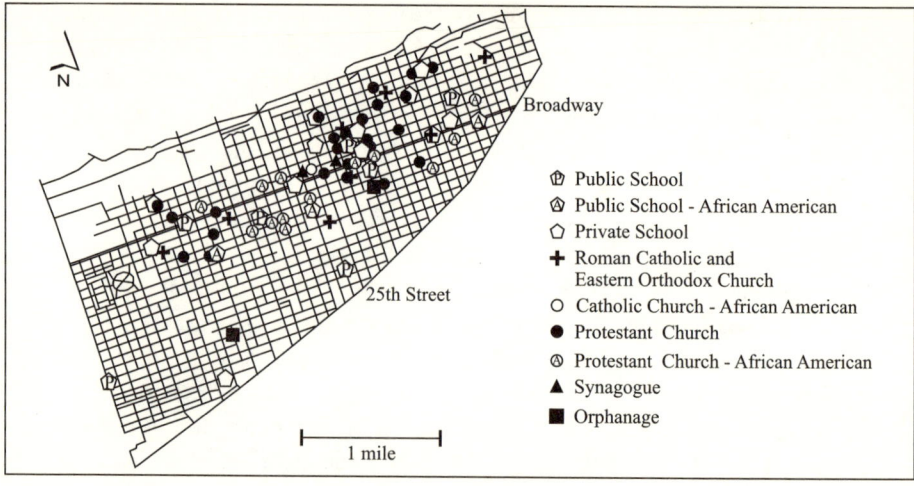

MAP 5.2. *Schools, churches, and orphanages in Galveston before the Great Storm. (Cartography by Linda F. Prosperie)*

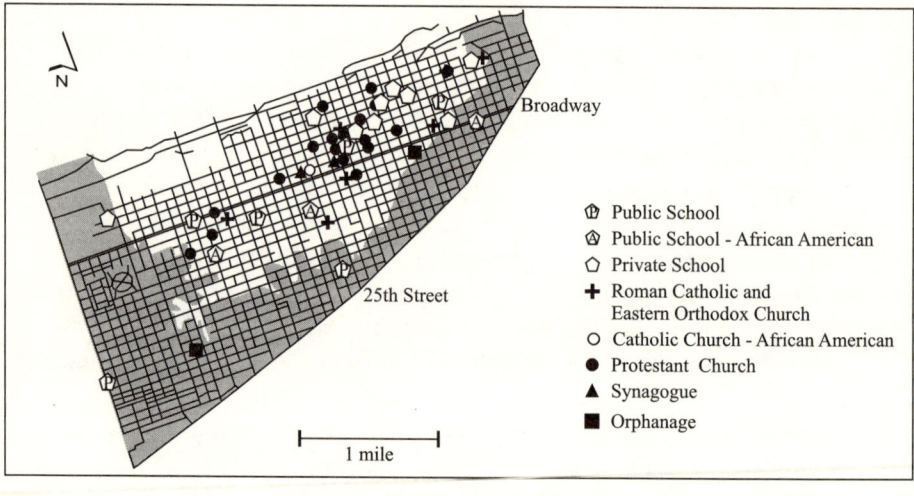

MAP 5.3. *School, churches, and orphanages in Galveston after the Great Storm. (Cartography by Linda F. Prosperie)*

TABLE 5.4. Public School Census, Galveston, 1900

	Students	Teachers
White schools		
Ball High School	631	31
Rosenberg	586	19
Second District	681	14
Third District	652	14
Fourth District	745	15
Fifth District	452	10
African American schools		
Central	231	6
East District	451	9
West District	641	21

ters supervising children in the asylum died in the storm, along with almost all of the children. Some of the nuns had tied children to their waists with long ropes in a vain effort to save them from washing away in the waves.

Galveston had a large number of school-age children, who were affected by the loss of both educational property and the stability of their daily school routines (table 5.4). Many other children attended private schools in the city and were likewise affected by the storm damage.

Virtually every school and orphanage in the city suffered damage from wind and water. Especially hard hit were all-black schools. The East District School, built in 1885, was destroyed. A local observer described this school the day after the storm: "It was a substantial wooden building equipped with all modern appliances and conveniences. Not a board or brick of this house can be found, so complete is its destruction" (Ousley 1900, 134).

THE EFFECTS OF THE STORM ON IMMIGRANTS: THE ITALIAN EXPERIENCE

More Italians than any other immigrant group lived in neighborhoods located on the Gulf side of the island at the time of the Great Storm (map 5.4). Italians, therefore, were more seriously affected by the damage of the storm because many lived within the zone of most intense storm damage. Most had lived on the island a very short time. Their homeland in Sicily

MAP 5.4. *Italian, Greek, and Chinese residential patterns before the Great Storm. (Cartography by Linda F. Prosperie)*

and other parts of Italy could not possibly have prepared them for what they experienced during the hurricane of 1900. Those who lived along the beach—if they survived the storm—witnessed the complete destruction of their homes and businesses. Italian immigrants who lived elsewhere on the island were shocked to discover after the storm that literally hundreds of their family members, close friends, and compatriots were dead or missing.

The child of two Italian immigrants, Lucy Rizzo, experienced the terrible loss of a parent during the storm. Lucy, age seven, and her mother were removed from their flood-wrecked house by men in a boat who were unable to return for her father. Unfortunately, he did not know how to swim. "She watched him with the wind and rain on her face for as long as she could, crying for him" (Macdonald n.d., n.p.). Mr. Rizzo was never seen again. At least seven other Italian children lived at St. Mary's Orphanage on the beach. None survived.

A comparison of the flood damage map (map 5.1) with the Italian residential map (map 5.5) reveals a close correlation between Italian residential patterns recorded in the manuscript census of 1900 and the areas most severely affected by the storm. Examination of these mapped patterns (and death records listed in the local paper throughout the week following the

AFTER THE STORM

hurricane) shows how severely this particular immigrant group suffered from the storm. In a strange coincidence, many newcomers from Italy who arrived in Galveston in the years immediately after the storm, such as grocer Arminio Cantini, supported themselves by working on the seawall, constructed to help protect the island's residents from future storms (Beasley 1999, 25). Two years after the storm Italian settlement patterns on Galveston Island had changed significantly, with many fewer Italians living near the beach (map 5.5).

A small number of Italian immigrants had settled in Galveston as early as 1857. According to listings published in the city directory that year, they opened small shops, operated ice cream parlors, and owned law firms. By 1880 the manuscript census listed the names and addresses of considerably

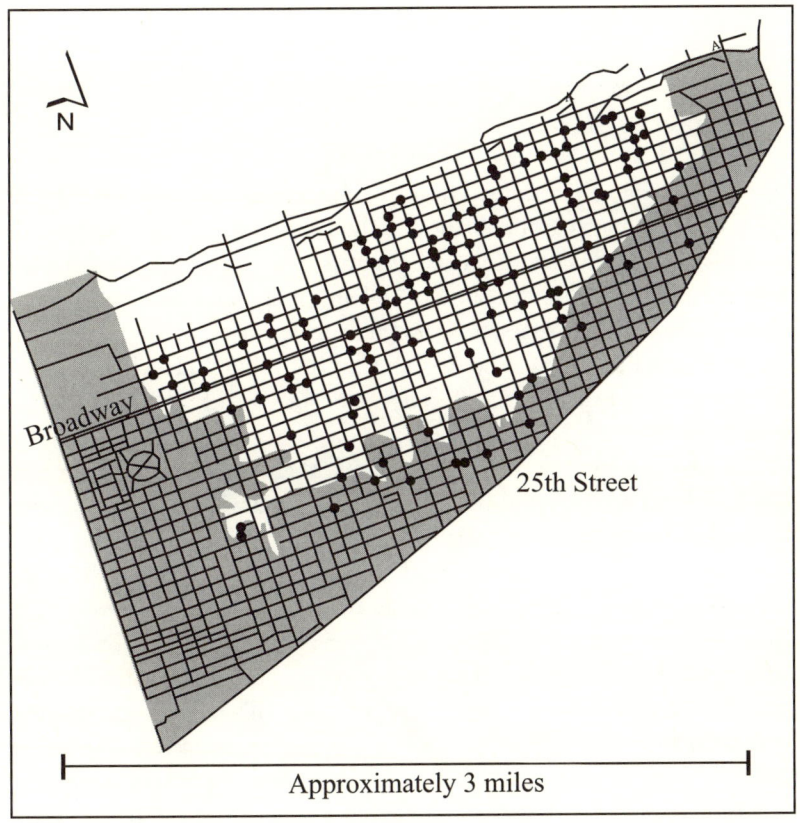

MAP 5.5. *Italian residential patterns after the Great Storm. (Cartography by Linda F. Prosperie)*

TABLE 5.5. Italian Immigrants Arriving in Texas, 1899-1910

Year	Northern Italians	Southern Italians
1899	114	96
1900	121	69
1901	130	111
1902	230	121
1903	227	151
1904	170	238
1905	183	239
1906	221	229
1907	287	284
1908	151	216
1909	176	122
1910	105	202
Total	2,115	2,078

Source: U.S. Congress, 1911b, 339.

more Italians living in the city. These late nineteenth-century Italian immigrants came primarily from northern Italy. They were joined by thousands of Sicilians and other southern Italians after the turn of the century. According to Belfiglio, "most southern Italian immigrants were poorly educated *pescatori* (fishermen), *contadini* (peasant farmers), or *giornalieri* (day laborers), who belonged to the Roman Catholic Church. Northern Italian immigrants were generally *uomini d'affari* (businessmen), who were Roman Catholic or Waldensians" (1989, 45).

One may assume that the fishermen and others who lived in beach-side neighborhoods on the Gulf side of the island—and who therefore suffered most from the storm—had come to Galveston from Sicily and other parts of southern Italy. They had followed a migration pattern established by a Sicilian fisherman who first migrated to the west cost of Florida and then relocated to Galveston. His decision in 1906 to move to the Texas Gulf Coast apparently launched a chain migration of southern Italian fishermen who moved to Galveston Island from the same village in Sicily. By 1906 more than 10 percent of all Italians living in Galveston were fishermen or were employed in port-related activities (Belfiglio 1983, 57; *Galveston City Directory* [1906], 324). The total number of southern and northern Italians who migrated to Texas between 1899 and 1910 is shown in table 5.5.

Other Italian immigrants who had been farmers in their Sicilian homeland arrived at the port of Galveston by way of other North American port cities, such as New York and New Orleans. Most of these agrarian newcomers, such as Pietro Maniscalco and his wife Maria Messina from the Sicilian village of Poggioreale in Sicily, stayed in Galveston only long enough to purchase a train ticket to the mainland (Marilyn Maniscalco Henley interview, 12 January 2000). Many settled in Dickinson and soon became successful strawberry farmers. Others traveled farther north to lease or buy land near Houston or plant crops along the fertile Brazos River floodplain (see Boykin 1993).

The Italian immigrants who stayed in Galveston added to the ever increasing membership of the three Catholic churches on the island: Sacred Heart, St. Patrick's, and St. Mary's (fig. 5.1). Most of the Italian Waldensians, who originated in the Tuscany region of Italy, joined the Presbyterian Church. By 1920 there were more than eight thousand foreign-born Italians living in Texas, mostly in the Galveston-Houston region, the Brazos Valley, and in Dallas-Fort Worth.

The building known as the Italian Vault provides evidence of the ongoing importance of Italians in the cultural landscape of Galveston. In 1992

FIGURE 5.1. *Sacred Heart Church is one of three Catholic churches that served Galveston's Italian immigrants and their families over the years. (Photo: Lee Miller)*

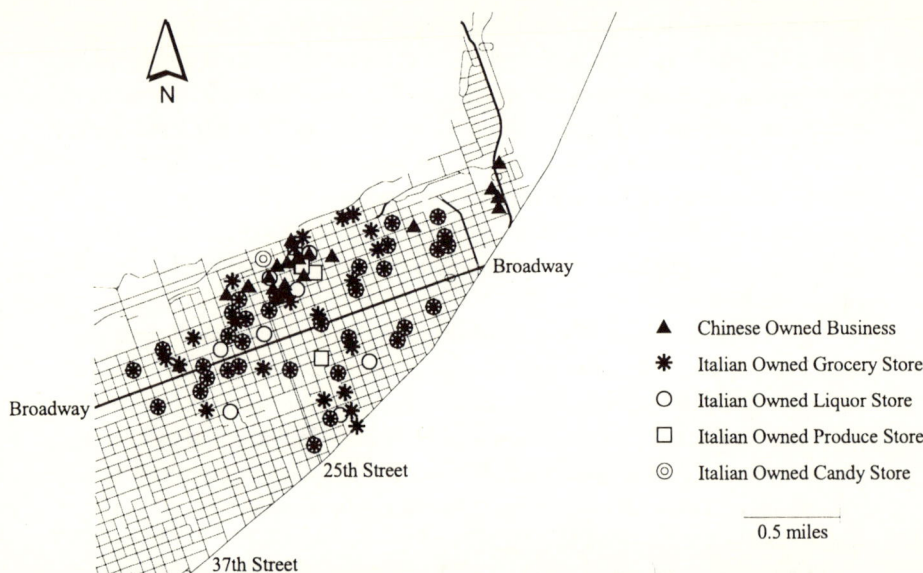

MAP 5.6. *Commerical ethnic landscape in Galveston, 1890s. (Cartography by Linda F. Prosperie)*

the San Giovanni Italian-American Association of Galveston dedicated the Italian Vault in a service held at the Calvary Cemetery on Sixty-fifth Street. This beautiful funereal building was constructed in 1888 by a group of the earliest Italians settling in Galveston. The original effort to construct the vault, to provide a final resting place for Galveston's less fortunate Italian residents, was led by the city's Italian consul, Clemente Nicolini, who came to Galveston from Italy in 1884.

Galveston's Italian community was definitely not dominated by the laboring class from its earliest days. Like Chinese and other occupation-niche immigrants in the city, most Italians were small-business owners. But while most Chinese residents of the island operated laundries, Italian entrepreneurs owned small groceries, fruit dealerships, and ice cream parlors (map 5.6). A count of all Italian-owned businesses listed in the city directory the year before the storm included 92 saloons, 79 groceries, 9 oyster dealers, and 9 fruit markets. There were also a number of clerks, salespeople, bookkeepers, and barbers (Belfiglio 1983, 55).

Beginning with the publication of *Il Messaggiero Italiano* in Galveston in 1906, Italian-language newspapers helped hold together Italian immigrants who lived in various parts of the region, state, and nation (table 5.6). They also helped maintain cultural ties and connections to the Italian homeland.

AFTER THE STORM

These publications—along with Italian festivals, church activities, family reunions, bocce ball tournaments, and mutual aid societies—helped maintain the integrity of Italian culture in Galveston and beyond. Many remain important today.

REBUILDING GALVESTON

Any traveler who has visited Galveston Island and who has also seen the now abandoned site of the former Texas Gulf Coast city of Indianola would wonder why the two places responded so differently to a deadly hurricane. Why did Galveston choose to rebuild after its devastation in 1900, while Indianola faded away after its last and worst hurricane in 1886? Why is Indianola now only a sandy beach, while Galveston has grown into a city of over sixty thousand people? How could two high-risk places in such similar locations have responded so differently to the threat of future hurricane damage?

Changes in City Government

Questions linger about decisions made in the aftermath of Galveston's Great Storm. Many residents at the time credited the reorganization of city government as the reason for Galveston's revival. Other Galvestonians believed the new commission form of government that was approved by city voters the year after the storm was simply another attempt by a few wealthy families to hold onto their control of island economic and cultural systems. The latter group insisted that "the movement for reform in municipal gov-

TABLE 5.6. **Italian Language Newspapers in Texas, 1914–63**

Newspaper	Place of Publication	Dates of Publication
La Tribuna Italiana	Dallas	1914–63
Il Messaggiero Italiano	Galveston and San Antonio	1906–14
La Patria degli Italiani	Galveston	1925
La Stella del Texas	Galveston	1913–18
L'Aurora	Houston	1906–19
L'America	Houston	1925
La Voce Patria	San Antonio	1925

Source: Marino 1981.

ernment ... constituted an attempt by upper-class, advanced professional and large business groups to take formal political power from the previously dominant lower- and middle-class elements so that they might advance their own conceptions of desirable public policy" (Hays as quoted in Rice 1977, xvi).

One thing is certain: The decisions of Galveston's all-powerful elite class —formerly played out through the Deep Water Commission and then through the new commission form of government—altered the island's physical environment as well as its human patterns.

Before the Great Storm, Galveston's system of city government was the same as those of most other cities of the era. Galveston was originally governed by a mayor elected by the votes of the entire city and twelve aldermen selected by residents in each of the city's twelve wards. The Texas legislature had changed this system by an act that called for sixteen aldermen, of whom four were to be elected at large and twelve by residents of the wards. This new alderman system created many problems in terms of fiscal management and related lobbying efforts of various aldermen who had been elected in two different ways by voters of the city (Cheesborough 1910, 4).

The decade before the storm had marked the beginning of efforts by financially successful entrepreneurs in Galveston to solve the inequities of ward-based politics in the city. New charter amendments in 1891 and 1893 provided for four additional councilmen to be elected at large. In 1895 the Chamber of Commerce recommended election of twelve council members, all to be elected by voters of the entire city. "Consequently the wards retained the appearance of direct representation but not the substance" because a ward candidate could be outvoted if he could not garner citywide support (Rice 1977, 4). This effectively ended officeholding by African American residents of the city, despite the large black population (22 percent of the city's total in 1895). Many African Americans were concentrated in one predominately black ward of the city (Hyman 1990, 156). This meant that even if an African American candidate could be elected by his own district, prejudice in the city as a whole would cause him to be outvoted (Rice 1977, 5).

After the storm Galveston was close to bankruptcy. The city infrastructure was in ruins. So too were its police and fire departments, educational institutions, churches, and most public buildings. Scrip was issued to cover for unavailable cash (Graham and Newman 1945, 89). It was obvious to those who had survived the storm that the city could not rebuild without financial help and complete reorganization of its government. The leaders of the most powerful political body in the city, the Galveston Deep Water Com-

AFTER THE STORM

mission, are credited with the development of the new form of government adopted after the storm. The Deep Water Commission had been established in 1882 to promote harbor improvements in Galveston. Its efforts had brought about the involvement of the federal government in "saving" Galveston's preeminence as the most important deep-water port in Texas through major sand removal and dredging projects (see chapter 4). Therefore, if devastation by the hurricane was not repaired, Galveston would lose its economic supremacy as the chief Texas Gulf port and the efforts of the Deep Water Commission would have been in vain. The group dominated three all-important sectors of Galveston's economy—banking, corporate management, and large property ownership—and so were in control of the city's destiny on a wide variety of economic and governmental fronts (Rice 1977, 7).

The Galveston Plan, as it was known, called for a board of municipal directors, to be composed of five members (a mayor-president and four commissioners), all to be elected at large by the qualified voters of the entire city every two years. Each commissioner was to be more than twenty-five years old and a citizen of the United States, and was to have lived in Galveston for at least five years immediately preceding appointment (Graham and Newman 1945, 91).

When this new system was presented to the Texas legislature, it was amended so that three commissioners would be appointed and two would be elected by Galveston voters. This system was tried for almost two years and then revised to its original intent of allowing all commissioners to be elected at large. The city was divided into five departments, each under the direction of one of the commissioners.

At first glance, this new form of government would seem to have represented all residents of the city equally. According to a report published by the Deep Water Commission in 1910, "this board is composed of five practical businessmen, each fully recognizing the fact that economy and business methods, not politics, should be employed in transacting the business affairs of the city" (Cheesborough 1910, 6). But an outsider who had experience with this same type of government in another city, had this to say about the system's potential for misuse of power: "The great danger to the elective commission form of government is that while at first it would be non-partisan, after a while the inevitable effects of partisan politics may appear and it will lose its high character. The ideal system is one under which the citizens treat all municipal matters as public business and not political business and that is not possible for any great length of time when they are divided into political parties by their opinions which have nothing to do with municipal business, and only confuse public opinion" (Cheesborough 1910, 9).

Good or bad, the Galveston Plan was adopted by over one hundred cities across the United States in the years after its development. By 1917 more than five hundred cities had adopted its provisions. Interestingly, 215 cities (only 4 percent of which had populations larger than five thousand) retained the plan by 1976 (Rice 1977, xiv). As one of the early thinkers who contributed to this type of city government concluded in 1912, "commission government is far from a perfect plan . . . it only marks a transition toward better things" (Childs 1912, 373).

How this new form of city government could spread so rapidly throughout the country and then lose importance so quickly has been the subject of numerous historical analyses and is beyond the scope of this chapter. However, without the impetus of the Great Storm in 1900, it is doubtful that Galveston, and hence other cities throughout the country, would ever have been motivated to develop and adopt this innovative approach to governing cities.

One of the most significant provisions of the new form of government in Galveston was empowering the board to appoint three competent and skilled engineers to devise a plan for elevating, filling, and grading the city for protection against the overflow waters of the Gulf during major storms. Another provision of the legislation authorized $1.5 million to cover the cost of these modifications (Graham and Newman 1945, 92). The new government, along with its plans to elevate the island to prevent further damage from hurricanes and other tropical storms, took effect in September 1901, a little over one year after the island-altering and life-changing impact of the Great Storm.

One might question how the lower- and middle-class citizens of Galveston, many of whom were new immigrants from southern and eastern Europe and African Americans, felt about the adoption of the new government structure in the city. But in the aftermath of the storm they faced survival issues of their own and so would have had no time for involvement in politics. Others no doubt believed that the commission form of government was the only way to secure state aid and maintain the local stability to repay bonds, rebuild the city, and protect it against future hurricanes. Still others may have trusted that this new approach would only be temporary until the storm damage was repaired. Finally, "the lower and middle-class public may have been simply willing to trust the judgment of the leading businessmen in time of acute crisis" (Rice 1977, 12).

AFTER THE STORM

Construction of the Seawall

Many projects had already been completed by local developers before 1900 with the aim of altering the natural environment of Galveston Island and its surrounding waters. Harbors had been dredged, sandbars moved, beach dunes removed and replaced by houses, and land filled in to provide additional space for industrial and commercial development. But none of these proved to have as dramatic and longterm an impact as projects carried out after the 1900 storm. Almost immediately after the massive destruction of the Great Storm, the island's political hierarchy embarked on an extremely ambitious plan to protect the people of the island and their investments. Decisions were made that would ultimately alter the barrier island—by nature a continuously changing type of landform—in an attempt to increase its stability. The construction of a great wall and plans to elevate the city so it could withstand the forces of nature were among the first two projects discussed. These major engineering feats forever changed the physical geography of Galveston Island.

As early as 1886 city leaders had discussed the possibility of constructing a seawall for protection from damaging storms. The idea at that time was considered far too expensive and much too difficult to implement.

The rationale for promoting the construction of the seawall and the grade elevation effort was simple. According to engineers hired by the city after the storm to study the problem, "the raising of the city grade is necessary to get the streets and lots sufficiently high for safety to life and property in severe storms . . . the sea wall is necessary to protect this filling from the force of the waves" (Walden 1990, 12).

The first step taken was data collection. Surviving citizens were asked to contribute information on the direction and force of ocean currents, the size and orientation of the debris wall, the water level before and after the storm, wind and wave contributions to the patterns of destruction, and whether or not there had been a current from the Bay to the Gulf or vice versa. It was discovered that the highest point in the city was along its main arterial street, Broadway. This street was 9 feet above mean low tide. The elevation for the remainder of the city averaged only 5.8 feet. Data provided by local residents affirmed that the storm surge measured at least 10 feet deep at Avenue A at Bayside and up to 15.7 feet deep in areas adjacent to the Gulf of Mexico. The storm surge had inundated the island from the eastern and southern ends. At some point, waters from the Gulf met with tides overflowing the island from Galveston Bay on the opposite side. This had

FIGURE 5.2. *Colorful murals on today's Galveston seawall obscure the reason for its existence—to protect the island from death and destruction by severe hurricanes. (Photo: Lee Miller)*

caused northwesterly currents to flow across the city throughout the storm (see Walden 1990).

Engineers hired by the city concluded that to protect the people from future storm events, the grade of the city should be raised and reconstructed to slope upward from the Bay toward the Gulf so that future storm surges would strike the highest side of the island. If Avenue A was raised to a height of 8 feet, and this elevation maintained from Avenue A inland toward the Gulf, the slope would be 1 foot per 1,500 feet.

The engineers also recommended construction of a seawall to protect the newly raised island. This concrete barrier was planned to be 3 miles long and 17 feet above mean low tide. The wall was to run in a southwesterly direction for twenty-three blocks. Its upper portion would be vertical to give the storm waves an upward direction and prevent water from flowing over the embankment behind the wall (Davis 1951, 5). The use of riprap along the front of the wall would provide added protection and prevent wave action from undermining its structural integrity. An embankment level with the top of the wall and sloping upward an additional foot for 200 feet behind the wall and then descending to the level of the city would preserve views of the Gulf and prevent water from flowing over the top during the worst storms.

When the city of Galveston finished constructing the wall as far as Thirty-ninth Street in 1904, another section was added, funded by the U.S. Congress to protect the Fort Crockett Military Reservation. In 1919 Congress authorized another extension from the east end of the wall at Sixth Street east to the Fort San Jacinto Reservation. This added another 10,300 feet of storm protection to the island. This east-end portion of the wall was extended even farther in 1923. Three years later Galveston County extended the west end of the wall from Fifty-third Street to Sixty-first Street, about three blocks from the site of the old St. Mary's Orphanage where more than one hundred people had lost their lives during the Great Storm. Figure 5.2 shows the seawall as it looked in 1999. Completion of this protective wall resulted in high-density commercial and residential development in the area behind it that now extends all the way across the island from the Gulf to the Bay.

FIGURE 5.3. *St. Patrick's Catholic Church in Galveston was only one of many public buildings raised above street level to protect it from potential damage from flooding caused by storms. (Courtesy of the Rosenberg Library, Galveston, Texas)*

FIGURE 5.4. *Huge pipes like these brought in sand from Bolivar Peninsula and beyond to help elevate the island city after the Great Storm. (Courtesy of the Rosenberg Library, Galveston, Texas)*

The process of raising the city began with the construction of levees around one small section of the city at a time. Every house, school, church, store, and any other structure in an area forty blocks long and two to twenty blocks wide was to be elevated. Additionally, all streets had to be repaved, which meant that all streetcar tracks, waterpipes, and other infrastructural support also had to raised. Buildings, including large masonry structures such as the church shown in figure 5.3, were lifted to the desired elevation by jacks. Residents were forced to walk on narrow planks when leaving and entering their homes and businesses. All flowers, shrubs, and trees had to be removed from the property if owners hoped to replant them when the project was completed (Graham and Newman 1945, 98).

After buildings within one area had been raised, sand soaked with water was pumped into the entire area (fig. 5.4). The water then drained back into the Gulf, leaving sand to settle to the top of each building's foundation. Fill depth averaged 5–20 feet. The comparative historic photographs of the level of the streets between Twenty-fifth and Twenty-sixth Streets in Galveston as measured on a telephone pole illustrate the impact of street raising on the city's infrastructure as well as its population (fig. 5.5).

The grade-raising project was not just local but actually global in scope. It served as the catalyst for an international technology transfer from Europe to the United States since the special dredges needed for the complex project had to be imported (Baker 1974, 8A). The first ship carrying dredging equipment for the sand and water arrived from Holland. Later two other ships arrived from Germany to handle the enormous job. Sand and other fill material was brought in from both the Gulf and Bay shorelines of the island and from Offatt's Bayou and Bolivar Peninsula located across the Bay due east of Galveston Island.

The total cost of the seawall and grade-raising projects was more than

FIGURE 5.5. *The two views of this house between Twenty-fifth and Twenty-sixth Streets in Galveston provide a vivid illustration of "before" and "after" the city's massive street-raising effort. (Courtesy of the Rosenberg Library, Galveston, Texas)*

$3.5 million. Approximately one-half of the cost was paid by local funds managed by the city government. The rest was supported by county, state, and federal funding (Graham and Newman 1945, 99–100). As mentioned earlier, the federal government had already spent a large amount of money to improve Galveston's harbor before the Great Storm and so had a vested interest in maintaining its earlier work. No doubt this prior fiscal investment, along with a familiarity and experience in modifying the natural environment of Galveston before the 1900 hurricane, laid the groundwork for federally funded dredging and other harbor improvements completed after the storm.

CONCLUSION

Engineering projects such as the seawall and grade-raising effort, along with harbor dredging and sand removal and rearrangement, irrevocably changed the physical geography of Galveston. But did they work? Was the island now safe from all future storms, and would it survive, unlike other Gulf Coast cities such as Indianola?

Galveston's major environmental modification schemes were severely tested in 1909 when another hurricane struck the Texas Gulf Coast about 45 miles southwest of Galveston Island. Water was thrown over the top of the seawall and drained across the fill. In addition, some of the riprap placed at the bottom of the wall was lowered by erosional power of the waves. As a result of this storm, repairs were made to the wall and the fill was extended to an elevation of 18 feet (Davis 1951, 7). Again in 1915 a major hurricane as strong as the 1900 storm hit the Texas coast only 30 miles southwest of Galveston. Despite severe damage to the seawall, only twelve lives were lost and property damage was considerably less than the damage incurred in 1900.

6

The End of Immigration

THE GALVESTON MOVEMENT

He was one among many to be rescued. In 1911 Gershom Geifman left his small farming village near the Ukrainian city of Kiev to travel to Galveston. The Galveston Movement had hoped to save them all. Geifman left home suddenly. He had no extra money to take and promised his new bride that after he had made enough money he would return and bring her back to America with him. Both were among the victims of the Russian pogroms which had persecuted Jewish farmers in Eastern Europe for decades.

Gershom Geifman's journey was long and confusing. First he had to make the painful break with his past and leave his village at Gorneshtople. Then came the 50-mile trip to Kiev, where he was helped by officials of the Jewish Territorial Organization (JTO). After leaving Kiev, young Gershom traveled by train west to Germany to the port city of Bremen where he finally boarded a steamship for the long voyage to Galveston. Gershom's brothers also were forced to leave their homes because of the extreme persecution of Jews by the Russian government. Five of them settled in the Quad Cities area of Iowa, one went to Mexico City, and another to Omaha. Gershom himself arrived in Galveston two weeks after leaving Bremen and was surprised and no doubt quite relieved to be welcomed by Rabbi Henry Cohen, local leader of Galveston's Jewish community, and Morris Waldman, a New Yorker hired by the Jewish Immigrants' Information Bureau to help welcome and resettle the newly arriving immigrants. They helped him and the other Russian passengers go through customs and complete their medical checkups at the Galveston docks. From there, Geifman almost immediately boarded the train north, where he started his new life working as a custodian at the John Deere factory in Rock Island, Iowa, part of today's Quad Cities area.

But this new immigrant soon was fired from his job at the factory because he couldn't write or speak fluent English. He spoke only Russian and Yiddish. Gershom then started selling brooms, which he carried on his back—in business for himself for the first time. Geifman's first economic breakthrough came when he saved enough money from the broom business to buy a blind horse. This enabled him to extend his sales territory, thereby increasing his earnings.

As promised, Geifman returned to his home village in 1913 to bring his bride to Galveston and then north to Iowa. When he finally got back home to pick her up for the trip, she had had a change of heart. Gershom then asked her younger sister if she would like to go with him and become his wife. She agreed. The story goes that their first son, Morris, was conceived on the boat.

The hopeful couple returned to the small home that Gershom had established for their new life in Rock Island, where he later worked his way up from selling brooms to collecting scrap metal from farmers, to buying and closing out failed grocery stores, eventually owning a large chain of eleven grocery stores in the area. When he died at the age of seventy-eight, Geifman was a respected and secure entrepreneur and investor and had a large number of real estate holdings.

It could have been a scene out of the popular musical *Fiddler on the Roof*. Over ten thousand Russian Jews were rescued from their rural villages in Russia and Ukraine and brought to Galveston in the early twentieth century. During this same time period and beginning more than half a century earlier, several million other European Jews, escaping persecution and economic crises in their homeland, had relocated to cities in the Northeast, the Midwest, and the West Coast. They made their new homes in New York, Philadelphia, and Pittsburgh; Chicago; or Los Angeles. These other, larger migration streams already have been documented and analyzed in many scholarly publications.

While this mass migration was taking place, a trickle of Jewish immigrants found their way into the United States by way of the Texas Gulf Coast via a well-organized plan that ultimately brought ten thousand of these rural Russian immigrants to the American South and Midwest. This effort, known as the Galveston Movement, has been little studied and remains virtually unknown among local residents.

Even among many Jewish historians and other scholars and writers in the United States, little awareness exists about the role of Galveston in bringing immigrants from Russia to this country. Perhaps no one has expressed this lack of understanding more vividly than writer Calvin Trillin (1996, 8–9):

> About all I knew of how my father's family got to St. Joe was that they went there directly from Galveston, Texas, where the boat from the Old Country had landed. When I was a child, I didn't realize that there was anything out of the ordinary in getting on a boat in darkest Europe, get-

ting off in Galveston, Texas, and going straight from there to St. Joseph, Missouri. Only later did it occur to me that what I had learned in school about the great wave of immigration from Southern and Eastern Europe at the turn of the century said nothing at all about the route my family had taken from suburban Kiev to St. Joseph, Missouri. . . . The Statue of Liberty was mentioned. The Lower East Side was mentioned. There was not a word about Galveston, Texas.

The timing of the Galveston movement was critical. Conditions in Russia were increasingly difficult, even life-threatening for residents of Russia who claimed Jewish heritage. And although the organizers did not know this, World War I was at hand—an event that would "seal the fate of an uncertain but vast number of thwarted potential beneficiaries of the Movement" (Hyman 1990, 243). In addition, the U.S. government was poised to pass a series of laws in the decade following the war that would severely limit the number of immigrants from Russia and elsewhere who would be allowed to enter the country. The seven critical years of the Galveston Movement were to be the final moments of mass immigration into Galveston, into Texas, and into the entire United States. It was also the first time any effort had been made in the United States to divert immigration from the East Coast to the territory west of the Mississippi River through a Gulf port (Shpall 1945, 119).

This complex effort involved globally minded Jewish leaders on both sides of the Atlantic. In Europe, the leading organizer was Israel Zangwill, the English playwright who coined the term *melting pot* in a play about a Russian Jewish immigrant escaping from his native land who falls in love with a beautiful Christian social worker. In New York it was Jacob Schiff, the American banker and philanthropist. And in Galveston Rabbi Henry Cohen, a leader of the Galveston Jewish community, and his local supporters, who were primarily wealthy German and Polish Jews, provided financial and social support for the effort. This team believed that their plan would help not only to spread Russian Jews throughout the midwestern and western states, but also to minimize the governmental pressure to pass new legislation to restrict further immigration into the United States.

PUSH FACTORS: THE SITUATION IN RUSSIA

Between 1903 and 1906 more than three hundred pogroms created panic among Jewish communities in Eastern Europe and especially in Russia, and forced large numbers of Russian Jewish immigrants to Palestine, South Africa, England, and the East Coast of the United States (Adler and Margalith 1946, 276). In 1901 a pivotal event occurred that shaped Jewish immigration rescue efforts in both the United States and Europe: the horrifying destruction of Kishinev, located in present-day Moldova. This massacre resulted in the death of forty-five Jews. Eighty-six others were seriously injured. Compared to later pogroms, this was a relatively small event, but it served as a catalyst for the movement to help those wishing to find new homes outside of Russia.

The pogroms that occurred in the early 1880s, and again just after the turn of the century as retribution for the role radical Jewish political leaders purportedly played in causing Russia to lose the Russo-Chinese War were only one part of the problem. Land reform, political turmoil and change, and economic crises had become a regular part of everyday life for most, making economic and political security at home difficult, if not impossible. As a direct result of these many overlapping issues, by 1891 Jewish organizations in the United States and Europe decided that emigration was the "only solution."

An additional problem facing Jewish residents of Russia was the strict geographical restrictions that confined residence to the twenty-five provinces in Russia and Russia-Poland known as the "Russian Pale of Jewish Settlement." Even inside this zone of restriction, the Russian government passed laws that permitted Jews to live only in specified rural areas, villages, and cities (Goldstein 1972, 15).

Several charitable organizations were organized in response to these pressing and often life-threatening needs. They provided support for new immigrants in a variety of ways from the time they left home to the time they reached their final destination. At first Jewish leaders in the United States, who were mostly German and relatively well educated, "were not deeply interested in the problems of Russian Jews and sometimes even resented the influx of the poor, illiterate, and culturally strange Russian immigrants" (Goldstein 1972, 12). According to Ornish, "these earliest Jews from England, Holland, France, Spain, and Germany were embarrassed at the sight of their coreligionists who had recently escaped the pogrom of Kishinev and other towns in Eastern Europe. The ragged, Yiddish-speaking Jews from Russia's outposts, ghettos, and *shtetls* appeared in marked contrast to the 'Our Crowd' Jewish families of New York" (1989, 121–22). But these same Jewish leaders, mostly New Yorkers, became concerned about the situation in Russia and assumed the obligation of organizing philanthropic aid for the victims of the pogroms and other forms of persecution.

THE END OF IMMIGRATION

PULL FACTORS: RUSSIAN IMMIGRATION INTO THE UNITED STATES

There were many reasons why Jews from Eastern Europe and Russia found the United States an acceptable alternative to their native land. The years following the Civil War were a time of great prosperity. Newcomers thus hoped to improve their economic situation by relocating to this more prosperous nation. Additionally, the significant Jewish presence already established in the United States meant that new immigrants could expect a warm welcome and could participate immediately in well-established traditions, social networks, and religious organizations.

An awareness of these benefits of seeking out a new life, coupled with the insoluble problems at home, prompted a mass immigration of Jews from Eastern Europe in the decades after the Civil War. Their numbers increased exponentially. Between 1870 and 1900 an average of 100,000 settled in the United States each year. The U.S. Jewish community grew from 250,000 in 1880 to over 3 million by 1914 (Bayme 1997, 350).

WHY GALVESTON?

Most American Jewish philanthropists had entered the United States at Ellis Island, along with tens of thousands of other immigrants. The vast majority of the new Jewish immigrants lived in densely settled neighborhoods on New York's Lower East Side, finding work in the garment industry or small shops. Most lived in crowded, congested flats, afraid to cross the boundaries of their language, religion, and culture. And there they stayed. "Often a family of six lived in one room with no running water, parents and children crowded together in one sleeping space" (Ornish 1989, 121). The more economically successful Jewish leaders in the United States not only worried about this situation, they were embarrassed by it. By 1905 there were almost one million Jews in New York alone. Many of the city's Jewish leaders and government officials strongly believed that something had to be done to redirect new arrivals away from New York toward the relatively less populated cities and states of the American Midwest.

The founder and funder of the Galveston Movement, Jacob Schiff, was one of these concerned leaders. After gaining the backing of the U.S. government, Schiff made contact with the Jewish Territorial Organization in Europe. The role of this London-based organization was to help select appropriate immigrants who could be easily employed and who were free of disease. If it selected a particular immigrant, the JTO purchased a ticket for that person for the trip from Bremen to Galveston.

At first Schiff favored the selection of the port of New Orleans, instead of Galveston, as an entry station for the newly arriving Jewish settlers. In a letter written in 1906 to Israel Zangwill, who headed the JTO, Schiff said:

> What I have in mind is that the Jewish Territorial Organization should take up a project through which it shall become possible to direct the flow of emigration from Russia to the Gulf ports of the United States—notably New Orleans—from where immigrants can readily be distributed over the interior of the country, I am quite certain, in very large numbers. From New Orleans, for instance, railroad lines diverge to the Pacific Coast, to the North and Northwest, as well as to the South, and Southwest, which provide easy and cheap transportation to these sections. After immigrants have once been landed at New York, Boston, Philadelphia, or Baltimore, they generally prefer to remain there, and notwithstanding all the efforts of the established removal offices only a comparatively small number leave these centers—aside from the great cost of transportation from the Atlantic cities to the Far West and Northwest, which makes the removal of large numbers from the East, where the great congestion prevails, to the great American "Hinterland" where a constant demand for labor of all kind exists, almost an impossibility. (Adler 1928, 97)

Schiff's vision, then, was to eventually provide for the resettlement of two million immigrants in underpopulated western states. Again in 1906 he wrote that "if properly handled and distributed [they] will be very welcome as well as useful to the country itself.... I believe that in government circles, as well as elsewhere, where the congestion of Russian immigrants in the Atlantic seaports is looked upon with a certain anxiety, such a movement will be regarded with great satisfaction and be given full moral support" (Szajkowski 1967, 24).

In the end, Galveston was selected over the port of New Orleans by both Jewish resettlement organizations for four primary reasons. First, Galveston's location was remote enough to divert Russians away from the well-known and much more popular port city of New York. Second, the North German Lloyd steamship line had direct contact with the port of Galveston. Third, the city was judged small enough that there would be little employment available for non–English-speaking newcomers, thus assuring that new arrivals would not decide to stay there permanently. Fourth, Galveston had railroad connections with the U.S. interior that extended as far north and west as the cities of Denver, Kansas City, St. Louis, and even Minneapolis. This would provide newcomers with transportation to cities that had been targeted for their long-term residency.

Arguments against port cities in other parts of the South were a bit more dramatic. Charleston, for example, was reported to be inhospitable to Jews and to encourage only Anglo-Saxon settlers. New Orleans was believed to be the worst place on the Gulf Coast for yellow fever and other epidemics. This largest of all Gulf Coast port cities also had a sordid and lurid reputation that made immigration organizers uncomfortable. They favored the city of Galveston, which numerous other European aid societies, particularly in Germany, had already used over the years as a port of entry for the resettlement of large farming communities in the interior of Texas. In the end, this successful precedent swayed the decisions of both Schiff and Zangwill in favor of Galveston.

THE JEWISH COMMUNITY IN GALVESTON

No doubt the existence of a stable Jewish community in Galveston was an important encouragement to officials making decisions about appropriate sites for the arrival of large groups of Jewish immigrants from Russia. As early as 1868 Galveston had at least a hundred Jews; by 1910, more than a thousand. In 1853 the city elected a Jewish mayor, Michael Seeligson, who had arrived in the city in 1838. Galveston thus had a Jewish mayor one hundred years earlier than New York City (Christian 1998, A1).

A couple named Osterman, who first arrived in Texas in 1839, were the guiding lights of the Galveston Jewish community in its earliest days. In 1852 it was Rosanna Osterman who urged the city to establish its first Jewish cemetery. This pioneer couple also helped change the landscape of Galveston Island by importing the first oleander plants from Jamaica in 1841 (fig. 6.1). They eventually opened a store downtown and later invited Rosanna's brother, Isidore Dyer, to join them. Dyer later became the head of the Galveston Jewish community. The first Jewish services on the island and in Texas were held on Yom Kippur in 1856 at the Dyer home in Galveston (Hewitt 1996, 5). In 1868 Temple B'nai Israel was founded. Another group, Congregation Beth Jacob, was chartered in 1930 by merging the island's two Orthodox congregations. Following many years of home services, the first synagogue on Galveston Island, now the Masonic Temple, was built in 1890 at 816 Twenty-second Street.

One of the many beneficiaries of the Ostermans' legacy was the Hebrew

FIGURE 6.1. *Tall oleanders like these bushes lining both sides of Broadway at the turn of the century were first planted by the city's earliest Jewish residents. (Courtesy of the Rosenberg Library, Galveston, Texas)*

Benevolent Society of Galveston, incorporated in 1866. This organization, like many related groups in the cities and towns of the United States and Canada, provided support for widows and orphans, organized dowries for brides, cared for the sick and indigent, and buried the dead. Less than a year and a half after Galveston's Hebrew Benevolent Society was formally incorporated, a devastating yellow fever epidemic struck. The epidemic filled the first Jewish cemetery with forty members of the Galveston Jewish community. Records maintained by the Hebrew Benevolent Society in Galveston indicate that Jewish residents of the island city "found it very difficult to endure the rigors of the Galveston climate," and that they "were very susceptible to the dread and often fatal scourge of yellow fever" (Dreyfus, Hebrew Benevolent Society, as quoted in Hyman 1990, 21). A second Jewish cemetery was established in 1868, and a third, in 1897. A fourth large plot of land to be used for the same purpose was purchased in 1951 (Hewitt 1996, 8–9). All are located roughly in the center of the island, halfway between Broadway and the open sea.

Other well-known Jewish residents in Galveston, such as the Kempners, are likewise direct descendants of early pioneers from Germany, Poland, and Russia. They too helped shape not only the Jewish community on the island but the economic and cultural destiny of the city itself. Perhaps none was better known than the activist Henry Cohen, who played such a major role in the successful years of the Galveston Movement.

Rabbi Cohen served the Galveston community for sixty-two years as a religious leader, philanthropist, humanitarian, scholar, social worker, and teacher. Born in London, he was first sent to serve the needs of the Jewish community at Kingston, Jamaica, and then Woodville, Mississippi, before being sent to serve the B'nai Israel congregation in Galveston in 1888 (fig. 6.2). His signature bow tie and dress jacket with coattails flying, his British accent, and his seemingly constant bicycle riding through the busy streets of the city's commercial and residential districts are remembered by many people today. Concerned for the needs of Jews living in remote parts of the state, Cohen regularly traveled to preach, bury, and perform marriage ceremonies for residents of towns from Nacogdoches in east Texas to Brownsville on the Rio Grande (Cohen and Cohen 1941, 74). Not only was Rabbi Cohen an active planner and player in bringing the ideas of the Galveston Movement to fruition, he also provided aid to the island's destitute residents after the Great Storm in 1900 and later, in 1914, provided support for Americans who were victims of the Mexican Revolution.

An active and ongoing social network was in place in Galveston when

FIGURE 6.2. *Temple B'nai Israel has been Galveston's largest and most active Jewish congregation for more than a century. (Photo: Lee Miller)*

the Galveston Movement began. As listed in the city directory for 1906–7, it included two synagogues, Congregation Ahavas Israel and Congregation B'nai Israel, and also a school, the Hebrew Orthodox School. There were a variety of Jewish social organizations as well:

> B'nai Zion Association
> Galveston Hebrew Orthodox Benevolent Society
> Hebrew Benevolent Society of Galveston
> Hebrew Ladies Benevolent Society
> Jewish Ladies Auxiliary Society
> Young Men's Hebrew Association
> Ladies Auxiliary to Young Men's Hebrew Association
> Zacharias Frankel Lodge No. 242

NETWORKS AND DESTINATIONS

The Galveston Movement was a part of a tightly organized network that spanned two continents. Potential immigrants were first contacted by the Jewish Emigration Society in Kiev through one of its eighty-two recruitment and placement committees throughout the Russian Pale of Jewish Settlement. Literature extolling the glories of the American West was distributed throughout this region. Once selected, individuals and families were given tickets for the train to Bremen, provided with lodging in this German port city, and handed letters of introduction, written in English, for their new employers in the United States.

The Bremen organization cabled the Jewish Immigrants' Information Bureau (JIIB) with a passenger list each time a ship left Germany. The JIIB manager in Galveston decided where to send the new arrival based on his employment skills. Meanwhile, Jewish leaders in midwestern cities and towns had been contacted about the pending arrival of new immigrants. Most of the new arrivals stayed in Galveston only a few days—or even a few hours—before leaving on the train for the interior. This gave them just enough time to go through processing, clear all paperwork, and receive information about their transportation to the north.

On site in Galveston, a young social worker named Morris Waldman was sent in 1907 to oversee the opening and operation of an immigration office to manage the affairs of the Galveston Movement. The JIIB Galveston office was opened soon thereafter. The organization rented an empty warehouse and remodeled it, adding comfortable furniture, a bathroom, and showers. It was to serve as a shelter and processing center for new immigrants. The shelter was completed and ready to open on a Saturday evening, even though its insurance coverage did not go into effect until the following Monday morning. At midnight, however, a fire broke out, and the whole building was destroyed. Waldman later found another building and started the on-site organizational effort all over again (Marinbach 1983, 15).

Once the JIIB office was back in operation after the fire, it began negotiating with the railroad in Galveston to cut fares for the new immigrants. Schiff dominated three of the largest railroad networks in the region, and he was willing to contribute $500,000 to the cause of redirecting new arrivals north and west. The JIIB accordingly made plans to send immigrants to selected places. They were to receive not only a ticket for the train ride north but also money to help them get started more easily in their new life.

Russian Jewish immigrants were placed in each city based on employment requests sent by letter to the New York office. Letters from El Paso, for example, expressed an interest in trunk, harness, and saddle makers; Corsicana asked for weavers, spinners, and doffers for its new textile industries; Waco needed shoemakers. The most frequently requested occupations were tailors, clerks, shoemakers, and carpenters (Ornish 1989, 123). Guidelines established by the network listed the types of immigrants who could be most easily absorbed according to their occupations. Easiest to place were strong laborers below the age of forty. Men who were ironworkers, carpenters, cabinetmakers, butchers, tinsmiths, painters, paperhangers, shoemakers, tailors, masons, plumbers, and machinists were especially encouraged to become part of the movement (Marinbach 1983, 14).

Zangwill had originally hoped to have the first contingent of immigrants leave for Galveston in early spring 1907. Arrangements had been made with the North German Lloyd Line to run steamers for this purpose all the way to Galveston with only one stop in Baltimore. The complexity of arranging for departures, employment in the United States, and financial support, however, delayed the sending of the first group until late June.

The ship that officially launched the Galveston Movement was the S.S. *Cassel*, which docked at the Galveston wharf on July 1, 1907. Figure 6.3 (*top*) shows passengers huddled together on the deck of one of the German ships and (*bottom*) the arrival of a typical ship carrying Russian Jewish immigrants to Galveston. The Galveston Movement passengers on the *Cassel*—and every passenger on every ship that followed it over the next eleven years—were met by Rabbi Cohen. Port officials then tagged, inspected, and approved (or did not approve) each passenger. The "Welcome to Texas" ceremony at the docks was a dramatic occasion:

FIGURE 6.3. *Ships arrived in the harbor at Galveston between 1907 and 1914, bringing in more than ten thousand Russian Jewish immigrants, along with other European newcomers. (Courtesy of the Rosenberg Library, Galveston, Texas)*

THE END OF IMMIGRATION

The "Cassel" docked according to schedule. The Rabbi immediately set to work collecting his charges from the steerage. He got them above deck as soon as possible, hurrying them along with friendly pats on the back, directing them in a fluent, staccato Yiddish . . . they had come from the steppes of Russia into the tropical warmth of a strange country. They were still dressed in woolen smocks; some were booted for snow, and a tall young fellow wore a high-crowned Russian hat. They huddled together on the aft deck—men, women, children . . . hot, Texas sun beat down on them; perspiration flowed in streams down their faces. (Cohen and Cohen 1941, 193-94)

After arrival at the port of Galveston, immigrants were distributed according to their occupations. Most were men who intended to send for their families after they secured work and found a place to live (Marinbach 1983, 96). Butchers went to stockyard cities, such as Kansas City, Fort Worth, and Omaha; carpenters, to the furniture centers like Grand Rapids and Topeka; and tanners, to Milwaukee. Jewish committees in each community took care of clothing and lodging for the new arrivals. Evening classes were established to enable Russian- and Yiddish-speaking Jews to learn English (Adler 1928, 103). Of this occupation-related migration stream, Trillin writes, "In order to disperse the immigrants . . . arrangements were made for jobs in various parts of the South and lower Midwest . . . my family had, in fact, gone to St. Joe specifically to work in the cabinet factory run by a German-Jewish family" (1996, 50).

Many of the Galveston Movement immigrants eventually went into business for themselves. Unable to get other jobs, many became self-employed by selling bananas, an exotic and unknown fruit that fascinated the Jews from Russia and Poland. Family memories of relatives beginning their new lives by selling bananas as peddlers and later becoming owners of dry-goods stores abound among descendants of Galveston Movement immigrants. According to Laurie Gudzikowski of the Institute of Texan Cultures in San Antonio, curator of an exhibit on Jewish Texans, "they would peddle the goods on their back until they got a little money, and then they would buy a horse, and when they got a little more money, they would buy a wagon . . . when they saved a little more money, they bought a store" (Perkes, 1998, A1).

In descending order, states receiving new immigrants from the Galveston Movement included Texas, Iowa, Missouri, Minnesota, Nebraska, Louisiana, Colorado, Illinois, Oklahoma, Kansas, Tennessee, Arkansas,

Wisconsin, North Dakota, and Mississippi. "Iowa received 1,225 immigrants, second only to Texas. Of the first thousand immigrants, one hundred went directly to Kansas City, Missouri. During the seven years of the Galveston Movement, 1,099 went to Missouri" (Ornish 1989, 123).

Jewish immigrants from Russia and other places in Eastern Europe had already migrated and settled in many of these same interior states in the years preceding the Galveston Movement. Texas, Kansas, Oklahoma, and Colorado were key destinations for Russian immigrants in the years between 1898 and 1904, much as they were during the Galveston Movement (map 6.1). This route relates closely to the existence of earlier Jewish communities in cities and towns in each of these states and elsewhere in the Midwest and West.

In most cases the Russian Jewish immigrants settled fairly smoothly into life in their new towns and cities in the interior of the United States. Most worked very hard to become part of the larger community while still maintaining the Jewish faith and culture. Most adjusted to their new culture quite well. Speaking of his own family, who were Galveston Movement immigrants, writer Calvin Trillin said,

> Our immediate family would not have struck anybody as foreigners. The fact that my father had been born in the Ukraine seemed almost a technicality. My parents had Midwestern accents, even though they both happened to be fluent in Yiddish. We didn't live in an immigrant neighborhood . . . but the Old Country—untalked about . . . was a constant in our lives. My mother was born in Kansas City, but her parents were immigrants. Her father, Pop, had one of those bone-chilling immigration stories so often heard among people of that generation. Having, according to his account, evaded conscription by two or three different armies, he sailed to American as a teenager, scheduled to be met at the boat by a half brother, but couldn't find him. He never found him. He never found anybody else he was related to. (1996, 48–49)

THE END OF THE GALVESTON MOVEMENT

The economic depression of 1907–9 was the first blow to be delivered to the big dreams of Galveston Movement organizers, who had envisioned rescuing at least two million people from brutal persecution in their homeland. This severe depression in the United States made finding employment difficult or impossible even for native-born residents of communities. Find-

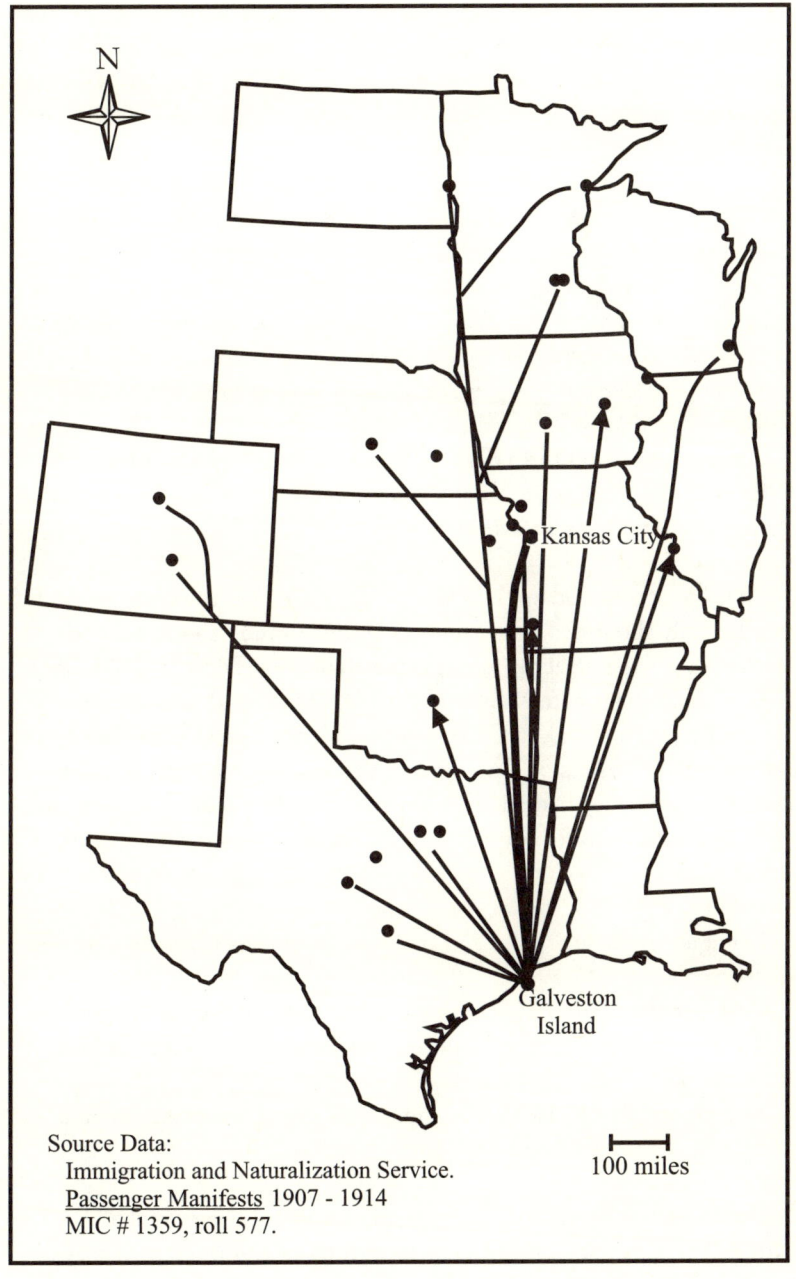

MAP 6.1. *The Galveston Movement: Migration streams of Russian Jews, 1907–14. Source: Immigration and Naturalization Service Passenger Manifests 1907–14. MIC #1359, roll 577. (Cartography by Linda F. Prosperie)*

ing and keeping full-time employment was even more difficult for non–English-speaking newcomers. Even the Yiddish press in the United States published complaints about continuation of the Galveston Movement in such dire economic times. In a letter to the editor, the director of the Kansas City Federation of Jewish Charities wrote:

> Everyone is employed at his respective trade, and with but one or two exceptions, earns at least nine dollars a week. Is it not gratifying, Mr. Editor, to think that right now, when thousands of people are out of work, when in Kansas City alone there are at least several thousand mechanics idle, that with a few exceptions, the Galvestonians should all be employed? . . . You know Mr. Editor, how very difficult it is to secure employment for some of the above mentioned class even under the most favorable circumstances. Yet, within three days of their arrival employment was secured for everyone of them. . . . what is true in Kansas City is likewise true, to an extent, in other places. (Shpall 1945, 140)

To try to accommodate these economic challenges, the Galveston office handled only about two thousand immigrants in 1909 and 1910. And these new arrivals had to meet some stricter stipulations. Most had to be heads of household, who agreed to send for their wives and children later. No one would be admitted at Galveston unless they had at least twenty-five dollars in their possession. Immigrants without skilled trades could not be admitted unless they had relatives to join in Texas or another western state. Ultimately new rules for admission were created by Zangwill's organization in London:

- Immigrants who are skilled and are under forty years of age must possess fifty rubles.
- Unskilled immigrants must show that they have relatives in this country.
- The immigrants must pay for the entire journey up to their place of destination.
- At the port of embarkation the immigrants must present the required amount of money (Shpall 1945, 156).

But the worst blow to the movement was the decision by the U.S. Department of Commerce and Labor to turn more and more people away at the port of Galveston and force them to return home—despite the severity of Russian retribution for returning emigrants. In the summer of 1910, for example, many Russian immigrants were accused of breaking the 1907 Immigration Law, which prohibited immigrants from having their passage paid by anyone but their own families. The government charged that the Galveston organizers had paid the transatlantic fares for some of the immigrants. Other new arrivals were said to be too poor to be anything but public charges.

In late 1910 a frustrated Jacob Schiff decided to take action. In a letter to the assistant secretary of the Department of Commerce and Labor, he protested the harsh treatment of Russian immigrants by Galveston port authorities:

> Since its beginning in 1907, the flow to Galveston, if not large, has been steady—a total of some two thousand having arrived there since the inception of the movement. These arrivals have consisted largely of men only who after gaining a firmer foothold, sent for their families, which latter often came through other ports than Galveston to join their breadwinner. I believe I do not go too far if I assert that almost none of those who have come through Galveston have become public charges, but, on the contrary, these immigrants have readily established themselves all over the West, Northwest, and Southwest, and have grown into useful additions to the working population of that section. . . . Of late, however, the Department of Commerce and Labor . . . is throwing needless difficulties in the way of the admission of those who arrived in Galveston, a course which, if persisted, is certain to break down the Galveston Movement. (Adler 1928, 105)

After 1913 port officials became even stricter and began to turn away an increasing number of Russians. In 1913 more than 5 percent were found ineligible for entry (Ornish 1989, 125). The largest number were rejected for medical reasons, because of hernias and vision problems. The percentage of rejections from Galveston docks became the highest of any port in the United States. Strangely enough, many of the immigrants from Russia who were rejected by Galveston port authorities were admitted to the United States at other places. According to Zangwill, writing in 1913, "if the conditions in Galveston remain as they are now, i.e., if the deportation of the selected immigrants from Galveston is seven times as large as that of Boston, six times that of Philadelphia, and four times that of New York, the situation is indeed very serious" (Shpall 1945, 157).

Female immigrants were especially vulnerable to deportation—especially young, single girls. For them, the only alternative to being sent back to Russia was finding immediate employment in Galveston (Marinbach 1983, 119). Rejection rates at the port of Galveston were especially high for female Russian Jewish immigrants. Four times more women were turned away by port authorities at Galveston (43 per thousand) than in New York. Jewish women were often denied admission between 1907 and 1913 because they were said to be "morally defective." Rumors in white, Protestant America circulated widely about the sins and dangers of both Catholic and Jewish newcomers. "The essence was . . . that crime flourished where Jews were, and that the importation of prostitutes, or, worse, the enticement of 'Americans' into the trade, was among the most lucrative of these highly organized crimes" (Hyman 1990, 246). Subsequently, many of the "exotic" Jewish immigrants who had entered the United States at Galveston were accused of everything: organized crime, prostitution, criminal trading schemes, and white slave traffic. From London, Zangwill expressed his concern to Schiff in a letter of December 5, 1913, and claimed that "if any of the women were prostitutes, they had joined the sisterhood of misery only after landing at Galveston!" (in Hyman 1990, 246).

By the end of 1912, despite all the problems caused by port authorities and the economic issues that made it difficult or impossible to assure newcomers full-time employment in destination cities and towns, more than five thousand new Jewish immigrants had been admitted at the port of Galveston. But as time went on, conditions in Russia deteriorated even more, hampering the rescue effort. Tales of problems with U.S. government authorities made the Galveston route lose its popularity among Russians who wished to relocate to North America. By 1914, a few months before the outbreak of World War I, the movement had come to an end. The subsequent death of Schiff and the negative attitude of the Yiddish press on both sides of the Atlantic, claiming that the Galveston Movement leaders "had failed to understand the Russo-Jewish immigrant and his aspirations," made revival of the movement at the end of the war an impossibility.

Did the Galveston Movement fail? Although it did not accomplish what it set out to do—to alter the directional flow of Jewish migration to North America—it did rescue at least ten thousand people and help them start new lives. In spite of the pressures brought about by an economic depression between 1907 and 1909 and the intense U.S. government deportation effort in 1910, descendants of those who were brought to the United States remember and appreciate the link with their homeland only because of the success of the Galveston Movement. The effort also helped in the fight against new restrictions on immigration into the United States and "evoked a new, militant political approach on the part of the American Jewish leadership" (Marinbach 1983, 181). These benefits, and the Jewish communities the Galveston Movement helped create and build in small towns and cities all across the American Midwest and West, remain as significant contributions even today.

The Reinvention of People and Place

At the turn of the century
A hundred years ago,
They came here on ships.
Fueled by burning dreams
They were sure would last.
Passengers from old Europe
And other places far,
Knew life would be best
Upon new land,
Holding on to the past.

The old lives left far behind
Felt callused and bruised up.
The ones up ahead
Promised and assured.
With memories in hand.
Housed vivid visions pulling
And pushing them along.
They treaded on hopes,
Kept dreams afloat,
Reaching that Texas sand.

The music that played in their hearts,
Accordions and bagpipes and such,
Soothed bitter times
While heartbreaking rhymes
Made laughter a much-needed touch.

They danced to the Galveston gait.
A tune that did surely inspire
Years of strong will

THE REINVENTION OF PEOPLE AND PLACE

For tough toil and till.
Galveston
Gateway to the Texas Empire.

With reasons in their pockets
Traveling in the darkness,
Holding their heads high
Above the waters.
Tumbling in the breeze
As their captains sailed onward
Cool dampness, their ally,
Never knowing what
Lay just ahead,
Minds were never at ease.

Descendants still remember
Tales that crossed the ocean,
Told to them in turn.
While the children dreamed on
About new life
Their new home, an old mirror
Shining at times brightly.
For some a reminder,
That coming here
Might be struggle and strife.

The music that played in their hearts,
Accordions and bagpipes and such,
Soothed bitter times
While heartbreaking rhymes
Made laughter a much-needed touch.

They danced to the Galveston gait.
A tune that did surely inspire
Years of strong will
For tough toil and till.
Galveston
Gateway to the Texas Empire!

"GALVESTON: GATEWAY TO THE TEXAS EMPIRE,"
LYRICS BY BETTO RAMIREZ AND GRACE
ARJONA-RAMIREZ, MUSIC BY LOS ROJOS DE ROMA, 1999

Are cities, like works of art, only viewed as successful if they survive the test of time? Does Galveston's existence today—despite the limitations of its meager resource base, the almost complete destruction by a major hurricane, cessation of population growth, and major economic decline in the mid-twentieth century—speak to its power as an urban center? Or do these indicators mean the city is doomed to ultimate failure? How did Galveston endure after the storm of 1900, when almost all infrastructure collapsed and a large percent of its population either died or emigrated? And how could this isolated "edge city" possibly have survived Houston's emergence as the wealthy "giant next door" in the early twentieth century? In the final analysis, has the geography of the island been its major raison d'être—or will it prove to be the ultimate reason for its decline?

The census of 1920 documents that the city of Galveston still had a diverse population at the end of World War I. This was to change as new U.S. immigration laws, establishing strict quotas for each immigrant group, were passed in the 1920s, effectively limiting the numbers of new arrivals at the port. Between 1920 and 1930 Houston grew 111 percent (an increase of 154,076 residents), while Galveston's population dropped by 2.1 percent. By 1930 Galveston had slipped to tenth position among Texas cities. Following World War II, during the era of growth elsewhere in Texas between 1950 and 1980, Galveston showed barely any increase in its population. Over the course of the twentieth century, the population of Galveston leveled off, while the populations of other large Texas cities grew by leaps and bounds between 1910 and 1990.

DECLINE OF POPULATION AND ECONOMIC POWER

What happened to the glorious dream of the golden isle? Galveston "was ordained to be the Seaport of the West, with a destiny of maritime ascendancy, of grandeur and of power" (Morrison 1890, 4). How could a place that was the state's largest and most architecturally beautiful city, and a major business node, up to 1890 undergo such a decline?

Contrary to popular belief, in the end the most dramatic and crushing blow to Galveston's growth was not the damage caused by the Great Storm. Like San Francisco after the 1906 earthquake or Chicago after the devastating fire that leveled much of the city in 1871, Galveston survived the disaster and rebuilt. Indeed, viewed from a development perspective, Galveston might be seen as the ultimate triumph of technology over nature: a remarkable success story of complete structural mitigation of the environment, a place where the efficacy of engineering genius truly saved the day.

Based on the data presented in this book, it appears that the decisions of a few elite families arbitrarily caused the demise of the booming city. The complete control of politics, the port, and most of the real estate on the island by these white, native-born, mostly southern businessmen meant that their decisions held firm no matter what the issue. Over and over again, these decisions leaned strongly to maintaining the city and their fortunes as "big fish in a small pond" (Barnstone 1966, 14). Describing the decline of Galveston, John Gunther wrote, "Three seigniorial families controlled it, and, to hold it within their grasp, deliberately sought to keep it from expanding and competing with Houston" (1960, 18).

Of particular significance in support of this argument is the control of Galveston's all-important port by the city's elite class. When these powerful leaders decided in 1869 to charge higher fees to shippers of oil, cotton, and other products of the hinterland, the largest shippers decided it was too costly to send their products through the port of Galveston. This decision, and the construction of a deep-water channel at Buffalo Bayou just outside of Houston, prompted these vital shipping companies to bypass Galveston altogether. Thus, because of the control of its economic and political sectors by the ruling class, the much smaller and more remote island city never had a chance. According to historian Earl Fornell,

> By following a shortsighted policy, the Islanders had forced the Mainlanders to take measures that were eventually to deprive Galveston of its greatest natural heritage. The leading Galvestonians were primarily mercantile-Southern gentlemen, who would have been at ease in the best society of Charleston or Philadelphia. Because they loved a conservative way of life they lost control of an empire to the "plungers of Houston" who, while they were perhaps as genteel as most 19th century Texans, were also rough-and-ready empire builders, who not only took advantage of the main chance but also literally created it with their own hands. (1956a, 1)

Human geography thus played the biggest role in the twentieth-century decline of the city of Galveston, rather than environmental devastation by the Great Storm, as many still believe. Like other towns and cities in much of the South—and despite its post-Enlightenment European connections—the city never evolved into a society where gender, race, ethnicity, and class divisions were reconciled. In short, despite its reputation as an open, accepting, and cosmopolitan place, its relatively integrated spatial patterns, and its ethnically and racially diverse population, Galveston is not and never was the egalitarian, multicultural "playground of opportunity" portrayed in much of the city's promotional literature.

In addition, although Galveston's offshore location may have been a real advantage for its growth in the mid-nineteenth century, when shipping dominated world trade and transportation, its small size and relative location put Galveston at a distinct disadvantage in the early twentieth century. While its geographical location initially favored the development of the city before the construction of a deep-water harbor in Houston, after the development of railroad networks in the interior of the state and nation Galveston lost out precisely because of its peripheral location on the edge of the state and continent.

SPATIAL PATTERNS VERSUS SOCIAL PROCESSES: A CITY IN CONFLICT?

One of the most fascinating outcomes of the mapped analysis of Galveston is the confusing incongruity of its spatial patterns and its social geography. If "ghettoization" of African Americans is the persistent geographic expression of Texas's racial dilemma and the most enduring feature of its cities (Davies 1986, 532), why did Galveston fail to fit this spatial model? If the city's social geography expresses the presence of uncrossable boundaries and "great divides" between racial and ethnic groups, why do its spatial patterns not indicate this arrangement? Specifically, why do the residential maps for the years 1857, 1880, and 1900 reveal such a surprisingly dispersed pattern for immigrant groups in the city?

The attempt to understand how a city like Galveston could be dominated by a white, southern-born elite class, while at the same time maintaining a highly integrated residential and commercial landscape, became the final quest in this study. The answers lie at the intersection of space and place, especially as these twin arenas focus on understanding ethnic identity.

As each of the geographically displaced immigrant and racial groups settled in Galveston, they almost immediately experienced the culturally integrating function of life in a densely settled and geographically isolated city. All identified themselves as German, Italian, "free black," Irish, and so on upon arrival in Galveston. As they gradually and sometimes quite painfully let go of their identity with Europe and places "over there" or "up there" to gain access to employment, these new residents became first Galvestonians and later Texans. But although they may have become Texans on the outside, they were still German or Italian or African American or Irish when they returned home from work, attended church, or visited with friends and family.

These multiple identities may have created an internal conflict that many never resolved. Diverse groups clung tightly together in religious and social networks that criss-crossed the island, linking residents of various parts of the city with each other. Because the residential districts in Galveston were small enough and close enough together to be easily crossed on foot, residents with the same ethnic or racial background were able to maintain close relationships with their ethnic identity and with each other no matter where they lived or worked—even though affordable housing was at a premium and had to be secured wherever it was available. The small size of their island habitat meant it was not necessary for particular groups to reside on the same street or even in the same neighborhood to stay closely connected in time and place.

The children of these first-generation Galvestonians no doubt were Texans in a deeper way, clinging only to their Old World identity as a piece of family history. These second-generation island residents seized on their status as "born on the island" (BOI) to safeguard their social position in island society. Many denied their identity as the descendants of immigrant families. Most became Galvestonians first and Texans second. As was true for other groups in other places, however, their children, especially those of the third generation who remained on the island, helped strengthen Galveston's immigrant connection and identity in the late twentieth century. This has contributed to the revival in recent decades of a potent "sense of ethnic place." The active expression of this cultural and ethnic resurgence is witnessed in the many activities of groups that work to keep out outsiders—those who will always be non-BOIs.

In the 1990s Galveston's residential patterns continued to be more integrated than those of other towns and cities in the South. Overall, on a scale of 1 to 100, Galveston's segregation index in 1990 measured 52, a score that reflects considerably more integration than in other nearby cities, such as Beaumont (segregation index of 62) and Port Arthur (65). Some variation in integration rates in different parts of the city of Galveston does occur (map 7.1).

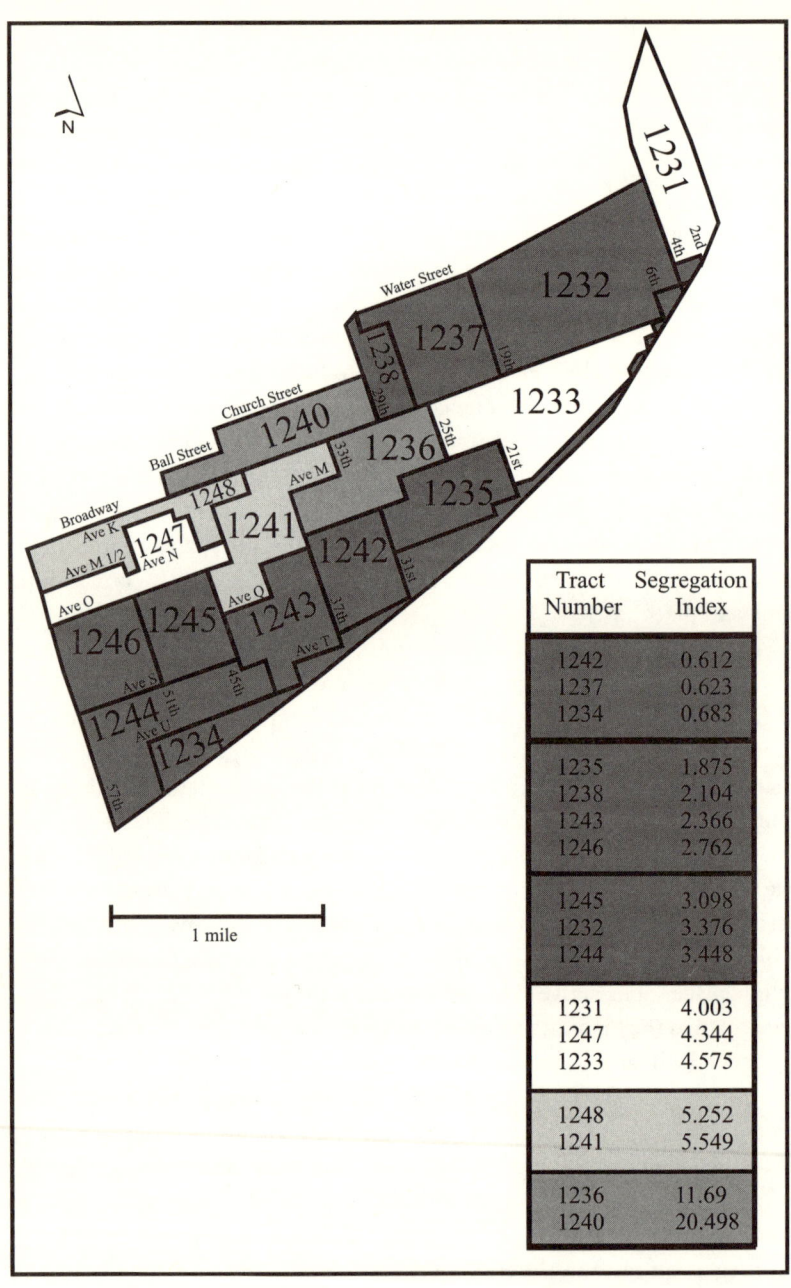

MAP 7.1. *Galveston segregation indices by census tract and district, 1990. (Cartography by Linda F. Prosperie)*

THE REINVENTION OF PEOPLE AND PLACE

FAILED ATTEMPTS AT REVITALIZATION

The twentieth century witnessed an ongoing effort by Galvestonians to recapture the city's former dominance as an urban center. In the 1920s the city reinvented itself as a destination resort, known primarily for the exciting night life of its movie-star visitors and its gambling casinos. Of particular importance during this period were a host of popular Italian-owned restaurants and clubs that opened downtown in the Strand district and along the seawall (fig. 7.1). By the late 1920s, however, it had become obvious that this niche would not be enough to attract population growth or to solve the city's financial problems in the long run. Something would have to be done to save the port.

By 1928 Galveston's competitive situation at the wharves had become so serious that a study of the port was undertaken and published in the *Galveston Daily News* (Cotner 1973, 136). It revealed that Galveston had priced itself out of the market with high dockage and storage fees, with higher rates on exports than on imports, and with ships charged by their total tonnage rather than by the goods they were carrying. As the economic situation in the state and nation worsened throughout the Depression in the 1930s, shippers began to depend more and more on the port of Houston.

The fishing industry was also tried as an avenue for economic growth. In the 1930s shrimpers in Galveston Bay proved to be so successful that Florida shrimp boats, which were larger and better equipped, began to move in and compete. Ultimately Texas and Florida shrimpers formed their own marketing association, known as the Gulf Coast Fishermen's Association, which contributed to Galveston's economy in a major way. One of the Florida shrimp dealers even opened a shrimp- and fish-packing plant in Galveston, aiding the city's economic recovery after the Depression (Cotner 1973, 138).

Galveston weathered the Depression better than most interior cities. It had been spared the trauma of bank failure, and its social service network was able to provide support for a 300 percent increase in the number of homeless people in the city's population. This was followed, however, by decades of high unemployment rates, loss of business at the port, and overall apathy by voters and activists about ever finding ways to solve the city's serious economic problems.

By the early 1940s Galveston's situation had become serious enough to warrant the formation of a housing authority, charged with ridding the city of its deteriorating low-income neighborhoods. Plans originally called for

FIGURE 7.1. *Gaido's restaurant, located on the seawall, has a clear view of the open Gulf. It remains one of the city's oldest and most popular Italian-owned eateries. (Photo: Susan E. Hume)*

the construction of 440 new low-rent apartments, 230 for African American residents of the city and 210 for white families. By the time the project was completed, however, the U.S. government had declared Galveston Island a critical housing area, and the authority was forced to rent all new apartments to war workers instead of to the city's often needy longtime residents (Housing Authority of Galveston, Texas, 1944, 4).

In the postwar decades, Galveston continued to struggle with economic problems and challenges brought by its diverse population. In 1956, for example, a Bi-Racial Committee on Desegregation published a report on educational issues in Galveston. It reported that "having studied the problem closely, and having full confidence in the ability of the school administration to administer the program in fairness and justice to every school child in Galveston, it is this Committee's recommendation that the principle of desegregation as decreed by the Supreme Court of the United States be applied to all schools and all grades in the Galveston Independent School District commencing in September, 1956" (Bi-Racial Committee on Desegregation 1956, 34). It took more than twelve years for the declaration of this committee to be achieved. In the fall of 1968 Galveston's schools were finally desegregated.

THE REINVENTION OF PEOPLE AND PLACE

REINVENTING THE CITY ONCE AGAIN

Beginning in earnest in the early 1970s, Galveston business and arts leaders turned their attention to preserving the city's historic buildings and districts. Building on a survey of the East End Historic District conducted in the late 1960s, the effort initially focused on the city's contemporary arts community. As in other cities such as Charlotte, New York, and San Francisco, arts leadership nurtured preservation work in Galveston. According to historic preservationist and author Ellen Beasley, who first came to Galveston in the mid-1970s when it was selected as one of three cities to be pilot projects for the American Bicentennial, "there had been relatively little work done in the neighborhoods, so most of the houses, a good many of them, were sort of unpainted white the same color as the sky . . . but the architectural historical resources here are just phenomenal and I'd just have to put it in a class . . . with a Savannah and a Mobile. It's an extraordinary city in that respect" (Beasley interview, 1980, 6, 9-10).

The Galveston Historical Society, established in 1871 by twenty-five Galveston businessmen to preserve city documents, is the oldest historical society in the state. More recent preservation efforts have been supported by the Galveston Historical Foundation (GHF), founded in the 1950s. This newer group was launched to purchase and save the Samuel May Williams home. After this preservation-minded foundation was created, the members of the Historical Society voted to dissolve the society and transfer their membership to the GHF, which now has more than forty-six hundred members, making it one of the largest locally based historic preservation efforts in the United States. In a city with more than 55 percent of housing serving residents of low or very low income, the foundation focuses on community redevelopment, public education, and historic preservation. Spinoff projects, such as the Texas Seaport Museum and the preservation of the historic sailing ship *Elissa*, have also contributed to the economic revitalization of the city by expanding recognition of the historic significance of Galveston's port (fig. 7.2).

The initial work of the GHF, preservation of the Williams home, set the stage for the preservation and redevelopment of the Strand commercial district, located one block from the harbor. In 1991 the GHF made a long-term commitment to revitalize the 1400 block of Church Street, an extremely deteriorated block in the heart of the East End Historic District. Four of the restored properties have now been sold to low- or moderate-income families that formerly rented property in the neighborhood. "Operation Church Street" continues today.

FIGURE 7.2. *The Texas Seaport Museum provides visitors and residents of the island with information about the importance of shipping at the city's busy international wharf. (Photo: Lee Miller)*

Although the usual local controversy has surrounded some of the decisions of the GHF over the years, its many successes cannot be denied. Not only have some of the city's most treasured historic properties been saved, but the economic development of the city has greatly improved. Attracting tourists to the city through annual events, such as the Historic Homes Tour, the Dickens-on-the-Strand festival, and the Sacred Places Tour, the GHF has clearly reenergized the tourism-based sectors of the city's economy in the past three decades. The National Trust for Historic Preservation presented the GHF with a National Preservation Award in October 1999, noting that "by preserving its past and building for its future, the City of Galveston has been transformed into a dynamic and historically rich community... the sheer scope and success of Galveston Historical Foundation's programs have made it a forceful model for local preservation organizations" (Kozel and French 1999, 1).

The GHF, along with the city's arts community, has provided leadership for other projects affecting Galveston's cultural landscape. Not all have been completely successful. The failure of one such project may be seen even by a passing motorist at the corner of Twenty-fifth Street and Broadway (fig. 7.3). A contentious and long-drawn-out struggle between the owners of Eckerd Drug Store, built on this section of Broadway long known for its

FIGURE 7.3. *Conflicts over appropriate land use, like the controversy over the Eckerd Drug parking lot on Broadway across the street from the "sacred" Texas Heroes Monument, have helped unify Galveston's various activist citizens' groups. (Photo: Susan E. Hume)*

historic Texas Heroes Monument and Sealy Mansion, and local attorneys, preservationists, and outside consultants took place in 1999. The central argument concerned whether to allow Eckerd to construct a large parking lot in the front of the building or to insist that it be located in a much less visible site behind the store. Eckerd and the developers eventually won out, despite an existing city ordinance that was to have protected Broadway from new development.

What of Galveston's future? Will the city's economic and cultural scene continue to be revitalized by historic preservation and related tourism efforts? Or will overdependence on the tourism industry cause the city's growth to "disappear with the next oil spill," as it has been said? Will renewed growth ultimately create the need for higher density development to support a much larger population on the island? If so, could this expansion bring new problems to an old city, such as increased traffic and crime? If development of the island continues to expand areally until the city reaches beyond the extent of the seawall to the west end of the island, will its dangerously low elevation and location on the open sea result in another hurricane event as destructive as the Great Storm of 1900?

These and other questions about Galveston's future remain unanswered. One thing is certain, however: Since most of the employment created by the tourism industry is at the level of minimum-wage jobs, the construction of new hotels and restaurants, and the expansion of tourist sites such as Moody Gardens, provides support for the same elite class system that dominated Galveston in its past. At least 60 percent of the city's population earns less than $23,000 a year. Forty percent of these families receive some type of government assistance. Young African American women make up 70 percent of the city's heads of household. Many local leaders, especially those in the African American community, assert that Galveston's tourism industry has made this situation worse instead of improving it (Valentine 1998, B1).

GALVESTON: PART OF A NEW NORTH AMERICAN REGION?

In chapter 1, I argued that the significance of a geographic analysis of Galveston was to provide grist for defining a new culture region rimming North America's Third Coast. My findings lend support to Jordan's contention that a unique region exists, located on the extreme edge of the continent. According to Jordan, this region may extend from Chesapeake Bay to the

THE REINVENTION OF PEOPLE AND PLACE

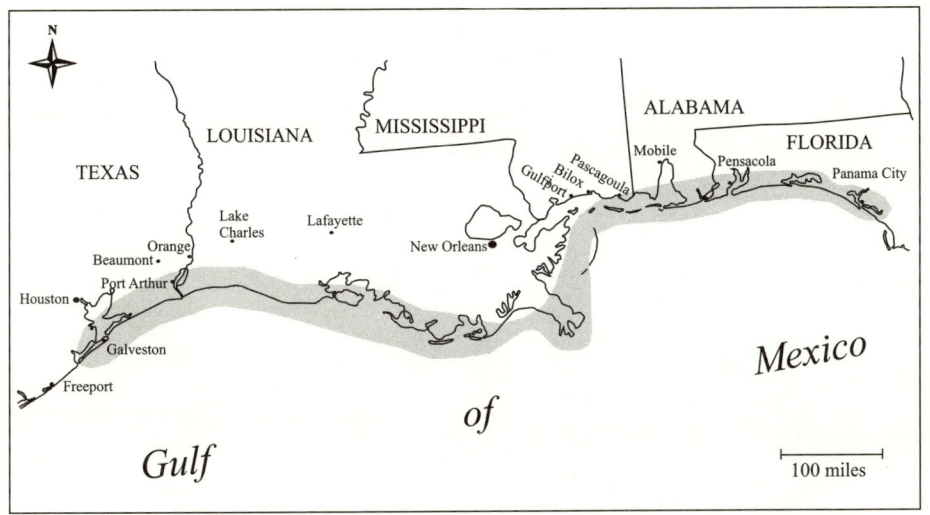

MAP 7.2. *Gulf Coast pivots: A region set apart? (Cartography by Linda F. Prosperie)*

Texas Gulf Coastal Bend. I have expanded the argument to apply to urban areas within this new region.

Jordan has called this place the "Creole Coast" (Jordan-Bychkov 2001). Set apart from the mainland physically, culturally, and economically, this Tidewater region "is more closely linked to the adjacent West Indian island world than to the rest of the United States" (ibid.). Jordan's argument is grounded in the idea that two intertwined economic and social systems prevailed in this region from the earliest colonial period: the slave plantation and cattle/hog herding. My emphasis has been on urban evolution and the creation of urban landscapes in the region, rather than on its rural socioeconomic and cultural systems.

The city of Galveston is one among a chain of small but essential urban areas nested within the Creole Coast (map 7.2). These cities have served as all-important hinges for the development and maintenance of this distinctive region, as they have alternately swung in and out over the course of the past two centuries. Cities like Galveston have provided essential connections both with the outside world and with the interior. This sets them apart. This also links each of them to the larger world beyond North America—even as it holds them fast to the less globally connected mainland.

FIGURE 7.4. *This storefront advertisement is a reminder of Galveston's lingering cultural connections to the outside world. The colorful image closely resembles artwork common in parts of the Caribbean and West Africa. (Photo: Susan E. Hume)*

The richness and longevity of Galveston's ethnic, racial, and cultural diversity reflect its position as a Gulf Coast pivot. Galveston is situated in both time and place. It is a place *apart from* but, at the same time, also a place *within*. The city's many links to the larger world—both past and present—may be seen in its colorful cultural landscapes (fig. 7.4).

THE REINVENTION OF PEOPLE AND PLACE

CONCLUSION

Further analysis of this larger Gulf region should add clarity to the often confusing, always complex story of the evolution of the city of Galveston. Its development, decline, and ongoing reinvention continue to prove fascinating. So too will related research themes in other cities located along the extreme southern edge of the American South. Scholars interested in larger issues relating to the evolution of space and the creation of place in other Gulf Coast pivot cities are encouraged to replicate and interrogate in other places the findings that have been presented in this book. Much work remains to be done by those interested in understanding more about the fluid intersections of history, culture, and economic development in this fascinating environment. I have provided here only the first pages of a much larger story.

BIBLIOGRAPHY

MANUSCRIPTS AND GOVERNMENT DOCUMENTS

All manuscripts and government documents were consulted in the Rosenberg Library Galveston and Texas History Center unless otherwise noted.

Addison, Oscar M. Letter to Mrs. I. S. Addison, 2 April 1845. Oscar M. Addison and Family Papers. Center for American History, Austin.
Alperin, Lynn M. 1977. *Custodians of the Coast: History of the United States Army Engineers of Galveston*. Galveston: Galveston District U.S. Army Corps of Engineers.
Armstrong, Neal F. 1987. *The Ecology of Open-Bay Bottoms of Texas: A Community Profile*. Washington, D.C.: U.S. Department of the Interior, U.S. Fish and Wildlife Service.
Babb, Stanley E. n.d. "A History of the Galveston Daily News." Unpublished MS [1969].
Beasley, Ellen. Transcript of interview, December 1980.
Bi-Racial Committee on Desegregation. 1956. Final Report on Educational Desegregation in the Galveston Public Schools. Report to the Board of Trustees of the Galveston Independent School District.
Bragg, Braxton. 1874. Letter to Mrs. Eliza Bragg, 13 July.
Charter Amendments and Revised Ordinances of the City of Galveston. 1855. Galveston: Civilian Book and Job Office.
Charter and Revised Ordinances of the City of Galveston. 1839, 1875, 1880, 1883, 1888, 1893, 1901–5, 1916, 1917.
Davis, Albert B., Jr. 1951. "Galveston's Bulwark against the Sea: History of the Galveston Seawall." Paper presented at the Second Annual Conference on Coastal Engineering, Houston.
Farrar, Roy Montgomery. 1926. *The Story of Buffalo Bayou and the Houston Ship Channel*. Houston: Chamber of Commerce.
Galveston County Tax Assessor Books. 1880–1914.
Galveston County Tax Rolls. 1838–1922.
Galveston Island: An Economic Profile. 1975. Galveston: Galveston Chamber of Commerce.

BIBLIOGRAPHY

Henley, Marilyn Maniscalco. Transcript of interview, 12 January 2000.

Housing Authority of Galveston, Texas. 1944. Public Housing in Galveston. Report of the Housing Authority of the City of Galveston, Texas, April 8, 1943, to April 8, 1944.

Immigrants and Passenger Arrivals. 1983. National Archives Trust Fund.

Index to Naturalization Records in Texas Courts. 1846-1939.

Kempner, Ruth Levy, Papers. 1961-63.

Konecny, Lawrence. 1995. *The Galveston-Bremen Project, 1865-1896*. Galveston: Self-published.

Mendoza, Carlos H. 1982. *Galveston Bay Area Navigation*. Washington, D.C.: U.S. Department of the Interior, U.S. Fish and Wildlife Service.

Morgan, Leon A. n.d. "The Impact of Desegregation on the Black Child." Unpublished MS.

Naturalization Records. U.S. Department of Justice, Houston.

Office of Immigration and Refugee Affairs. 1997. *Immigrants in Texas*. Austin: Texas Dept. of Human Services.

Passenger Manifests. 1847-1920. U.S. Immigration and Naturalization Service, Houston.

Port of Galveston Records. 1846-1950.

Report on the Improvement of the Entrance to Galveston Harbour and the Commercial Results Therefrom. 1898. Diplomatic and Consular Reports No. 464, Miscellaneous Series. London: Government Printing Office.

Rice, Hugh. 1867. *Report on the Survey of Buffalo Bayou, San Jacinto River, and Galveston Bay from the City of Houston to the Gulf of Mexico*. Houston: Gray, Smallwood and Co.

Scroll Project Records. 1984-92. Texas Jewish Historical Society. Rosenberg Library, Galveston. Processed by Anna B. Peebler, October 1997.

Scull, Ralph Albert. n.d. "Black Galveston: A Personal View of Community History in Many Categories of Life." Unpublished MS.

Ship's Passenger Lists, Port of Galveston. 1846-71. Galveston Genealogical Society.

Smith, Ashbel, Papers. 1839.

Statement of the Financial Condition of the City of Galveston, Texas. 1873. Galveston: News Steam Job Office. 28 February.

Truehart, Henry M., Papers. 1866.

U.S. Army Corps of Engineers. 1901. *Reports of Examination and Survey of Galveston Bay, Texas*. Washington, D.C.

U.S. Army Special Board of Engineers. 1890. *Deep Water at Galveston*. Senate Miscellaneous Document 89. 51st Cong., 1st sess.

U.S. Census of Population, City of Galveston. 1850-1920. Manuscript.

U.S. Congress. 1911a. *Reports of the Immigration Commission, 1911. Statements and Recommendations Submitted by Societies and Organizations Interested in the Subject of Immigration*. Senate Document 764. 61st Cong., 3d sess. Washington, D.C.: Government Printing Office.

BIBLIOGRAPHY

U.S. Congress. 1911b. *Reports of the Immigration Commission. Statistical Review of Immigration 1820–1910. Distribution of Immigrants 1850–1900*. Senate Document 756, 61st Cong., 3d sess. Washington, D.C.: Government Printing Office.

U.S. Department of Agriculture, Soil Conservation Service. 1988. *Soil Survey of Galveston County, Texas*. Washington, D.C.

VITAL STATISTICS

Customs House Records of the Port of Galveston. 1837–46. Compiled by Gail Borden.

Death Certificates, City of Galveston. c. 1880–1910.

Directory of the City of Galveston, 1909–1910. 1909. Galveston: Morrison and Fourmy.

Fayman and Reilly's Galveston City Directory for 1875–6. 1875. Galveston: Strickland and Clarke.

Galveston City Directory. 1906. Galveston: News Printing Office.

Galveston City Directory for 1856–57. 1856. Galveston: News Book and Job Office.

Galveston Directory for 1859–60. 1860. Galveston: Real Property Records.

"Handbook of Texas Online." 1999. Austin: Texas State Historical Association.

Hokanson, Joel. 1991. "The Immigration Database." Galveston: Galveston Historical Society.

Morrison and Fourmy Directory Company's Galveston City Directory, 1919. 1919. Houston: Morrison and Fourmy.

Morrison and Fourmy's General Directory of the City of Galveston, 1888–89. 1888. Galveston: Morrison and Fourmy.

Morrison and Fourmy's General Directory of the City of Galveston, 1895–96. 1895. Galveston: Morrison and Fourmy.

Morrison and Fourmy's General Directory of the City of Galveston, 1901–1902. 1901. Galveston: Morrison and Fourmy.

Mortuary Records, City of Galveston. 1875–1926. Book 1, 1 January 1875–31 December 1886. Book 2, 1 January 1887–9 August 1894. Book 3, 10 August 1894–30 March 1900. Book 4, 30 March 1900–31 December 1903. Book 5, 1 January 1904–31 January 1910. Book 6, 1 February 1910–31 May 1918. Book 7, 1 June 1918–31 December 1926.

Record of Internments, City of Galveston. 1859–72. Indexed and published by Peggy Gregory.

Register of Births, City of Galveston. 1910–31. Book 1, 1 February 1910–30 September 1915. Book 2, 1 October 1915–31 August 1921. Book 3, 1 September 1921–31 August 1926. Book 4, 1 September 1926–July 1931.

EDUCATIONAL RECORDS

Galveston School District Records.
Ball High School, 1886–92.

First District School, 1884-92.
Grammar School, 1881-88.
Second District School, 1881-84.
West District Colored School, 1885-91, 1892-1902.
Galveston Scholastic Census. 1900-1912.

SECONDARY SOURCES

Adler, Cyrus. 1928. *Jacob H. Schiff, His Life and Letters.* 2 vols. New York: Doubleday.

Adler, Cyrus, and Aaron M. Margalith. 1946. *With Firmness in the Right: American Diplomatic Action Affecting Jews, 1840-1945.* New York: American Jewish Committee.

Adler, Frank J. 1972. *Roots in a Moving Stream: The Centennial History of Congregation B'nai Jehuda of Kansas City, 1870-1970.* Kansas City, Mo.: Kansas City Printers.

Allen, Ruth. 1941. *Chapters in the History of Organized Labor in Texas.* Austin: University of Texas.

Axelrod, Bernard. 1966. "Galveston: Denver's Deep Water Port." *Southwestern Historical Quarterly* 20:217-28.

Baker, T. Lindsay. 1974. "Galveston Groundbreaking Was a Big Engineering Project." *Galveston Daily News,* 11 May, 8A.

———. 1982. *The Polish Texans.* San Antonio: Institute for Texan Cultures.

Barker, Eugene Campbell. 1902. "The African Slave Trade in Texas." *Quarterly of the Texas State Historical Association* 6:145-58.

———. 1918. *A School History of Texas.* Chicago: Charles Shirley Potts.

Barker, Eugene C., ed. 1929. *Readings in Texas History for High Schools and Colleges.* Dallas: Southwest Press.

Barnstone, Howard. 1966. *The Galveston That Was.* New York: Macmillan; Houston: Museum of Fine Arts.

Barr, Alwyn. 1996. *Black Texans.* 2nd ed. Norman: University of Oklahoma Press.

Barr, Alwyn, and Robert A. Calvert, eds. 1997. *Black Leaders: Texans for Their Times.* Austin: Texas State Historical Association.

Barratt, Ernest S., and Leon A. Morgan. n.d. "The Effects of Racial Integration on Academic Performance and Attitudes among High School Males: A Three-Year Study." Galveston: Self-published.

Baughman, James P. 1962. "The Maritime and Railroad Interests of Charles Morgan, 1837-1885: A History of the 'Morgan Line.'" Ph.D. diss., Tulane University.

Bayme, Steven. 1997. *Understanding Jewish History: Texts and Commentaries.* KTAV Publishing House and American Jewish Committee.

Beasley, Ellen. 1996. *The Alleys and Back Buildings of Galveston: An Architectural and Social History.* Houston: Rice University Press.

———. 1999. *The Corner Store: An American Tradition Galveston Style.* Washington, D.C.: National Building Museum.

BIBLIOGRAPHY

Belfiglio, Valentine J. 1983. *Italian Experience in Texas.* Austin: Eakin Press.

———. 1989. "The Nature and Impact of Italian Culture upon Galveston Island." *East Texas Historical Journal* 27:44-54.

Best, Gary Dean. 1978. "Jacob H. Schiff's Galveston Movement: An Experience in Immigration Deflection, 1907-1914." *American Jewish Archives* 30, no. 1:43-79.

Blasi, Edward J. 1965. "The Rise and Fall of Community Government in Galveston." Thesis, St. Mary's University.

Blasig, Anne. 1954. *The Wends of Texas.* Brownsville: Springman-King Printing. Reprint, 1979.

Bolin, Robert C., and Patricia A. Bolton. 1986. *Race, Religion, and Ethnicity in Disaster Recovery.* Boulder: Institute of Behavioral Science, University of Colorado.

Bolton, Herbert Eugene. 1915. *Texas in the Middle Eighteenth Century.* Berkeley: University of California Press.

Boykin, Rosemary DePasquale. 1993. *The Italians of Steele's Store, Texas.* Nacogdoches: Ericson Books.

Brashear, Etta A. 1941. "Galveston: Past, Present, and Future." M.A. thesis, Sam Houston State Teacher's College.

Breck, Allen du Pont. 1960. *The Centennial History of the Jews of Colorado, 1859-1959.* History Dept. Series, "The West in American History," 1. Denver: University of Denver, Hirschfield Press.

Bressler, David M. 1910. *The Removal Work, Including Galveston.* Proceedings of the National Conference of Jewish Charities.

Buchorn, Anne L. Moore. 1962. "The Yellow Fever Epidemic of 1867 in Galveston." M.A. thesis, University of Houston.

Campbell, Randolph B. 1989. *An Empire for Slavery: The Peculiar Institution in Texas.* Baton Rouge: Louisiana State University Press.

Cannatella, Mary M., and Rita E. Arnold. 1985. *Plants of the Texas Shore: A Beachcomber's Guide.* Sea Grant Program. College Station: Texas A and M University Press.

Carroll, James. 1905. *Yellow Fever: A Popular Lecture.* Austin: University of Texas Press.

Cartwright, Gary. 1991. *Galveston: A History of the Island.* Fort Worth: Texas Christian University Press.

Casteneda, Terri Alford. 1993. "Preservation and the Cultural Politics of the Past on Historic Galveston Island." Ph.D. diss., Rice University.

Cheesborough, E. R. 1910. *Galveston's Community Form of City Government: Its History, Details, and Practical Workings.* Galveston: Galveston Deep Water Committee. Reprinted from *Galveston Tribune,* 31 December 1909.

Chesnutt, Charles B., and Robert F. Schiller, Jr. 1971. *Scour of Simulated Gulf Coast Sand Beaches due to Wave Action in Front of Sea Walls and Dune Barriers.* College Station: Texas A and M University Press.

Childs, Richard S. 1912. "What a Democracy Should Be Like." *Everybody's* 26:373.

Christian, Carol. 1998. "Living History: Synagogue Served Island for 130 Years." *Galveston Daily News,* 26 March, A1.

BIBLIOGRAPHY

Claflin, James M. 1985. "Two Communities, Two Responses: An Alternative Interpretation of Calamities Applied to Indianola and Galveston." M.A. thesis, University of Texas, Austin.

Cohen, Nathan, and Anne Cohen. 1941. *The Man Who Stayed in Texas*. New York: Whittlesey House.

Cotner, Robert C., et al. 1973. *Texas Cities and the Great Depression*. Austin: Texas Memorial Museum.

Crouch, Barry A. 1992. *The Freedman's Bureau and Black Texans*. Austin: University of Texas Press.

Cuney-Hare, Maud. 1913. *Norris Wright Cuney: A Tribune of the Black People*. New York: Crisis. Reprint, New York: G. K. Hall and Co., 1995.

Davies, Christopher S. 1986. "Life at the Edge: Urban and Industrial Evolution of Texas, Frontier Wilderness—Frontier Space, 1836-1986." *Southwestern Historical Quarterly* 89:443-554.

Davis, Kathleen. 1970. "Year of Crucifixion: Galveston, Texas." *Texana* 13:141-54.

Dielmann, Henry B. 1960. "Emma Altgelt's Sketches of Life in Texas." *Southwestern Historical Quarterly* 63:363-84.

Doughty, Robin W. 1983. *Wildlife and Man in Texas: Environmental Change and Conservation*. College Station: Texas A and M University Press.

———. 1987. *At Home in Texas: Early Views of the Land*. College Station: Texas A and M University Press.

Dowell, Greensville. 1876. *Yellow Fever and Malarial Diseases, Embracing a History of the Epidemics of Yellow Fever in Texas, New Views on its Diagnosis, Treatment, Propagation, and Control*. Philadelphia: Medical Publication Office.

Dreyfus, Stanley A. 1963. *Henry Cohen, Messenger of the Lord: A Tribute to the Memory of His Beloved Rabbi on the 100th Anniversary of His Birth by Congregation B'Nai Israel of Galveston*. New York: Bloch.

Dugas, Vera Lea. "A Duel with Railroads: Houston vs. Galveston, 1866-1881." *East Texas Historical Journal* 2:118-25.

Eisenhour, Virginia. 1983. *Galveston: A Different Place*. Galveston: By the author.

Faye, Stanley. 1931. "Trouble on the Coast." *Southwest Review* 16:469-83.

Fiorenza, Joseph A. 1992. *Italian Vault: 1888*. Galveston: San Giovanni Italian-American Association of Galveston.

Flannery, John Brendan. 1995. *The Irish Texans*. San Antonio: Institute for Texan Cultures.

Fleishaker, Oscar. 1957. "The Illinois-Iowa Jewish Community on the Banks of the Mississippi River." Ph.D. diss., Yeshiva University.

Fornell, Earl W. 1955. "Ferdinand Flake: German Pioneer Journalist of the Southwest." *American-German Review* 21:25-28.

———. 1956a. "Galveston-Houston Rivalry." *Houston Post*, 9 May 1956, 1-2.

———. 1956b. "The German Pioneers of Galveston Island." *American-German Review* (February-March): 15-17.

BIBLIOGRAPHY

———. 1961. *The Galveston Era: The Texas Crescent on the Eve of Secession.* Austin: University of Texas Press.
Frantz, Joe B. 1988. *Lure of the Land: Texas County Maps and the History of Settlement.* College Station: Texas A and M University Press.
Fuller, Franklin R. IV. 1990. "Galveston Bay: A Study of State and Federal Policy." M.P.A. thesis, Lyndon Baines Johnson School of Public Affairs, University of Texas, Austin.
Galveston Daily News. 1942. Centennial Edition, April.
Gamblin, William W. 1993. "Estimation of the Contribution of Streams toward Material Loading to the Galveston Bay System." M.S. thesis, University of Texas, Austin.
Garner, Ruby. 1927. "Galveston during the Civil War." M.A. thesis, University of Texas, Austin.
Gatschet, Albert S. 1891. "The Karankawa Indians: The Coast People of Texas." *Archaeological and Ethnological Papers of the Peabody Museum* 1.
Giles, Bob. 1991. "'Italian Vault' Blessed, Dedicated in Ceremonies Held in Galveston." *Texas Catholic Herald,* 9 August, 1-2.
Goldstein, Judith. 1972. "The Politics of Ethnic Pressure: The American Jewish Committee as Lobbyist, 1906-1917." Ph.D. diss., Columbia University.
Graham, Sam B., and Ellen Newman. 1945. *Galveston Community Book: A Historical and Biographical Record of Galveston and Galveston County.* Galveston: Arthur H. Cawston Publisher.
Green, Casey Edward, and Shelly Henley Kelly. 2000. *Through a Night of Horrors.* College Station: Texas A&M University Press.
Guderjan, Thomas H., and Carol S. Canty. 1989. *The Indian Texans.* San Antonio: Institute of Texan Cultures, University of Texas.
Gunn, Clare A. 1969. *Annotated Bibliography of Resource Use: Texas Gulf Coast.* College Station: Texas A and M University Press, Sea Grant Program.
Gunther, John. 1960. *Taken at the Flood: The Story of Albert D. Lasker.* New York: Harper and Brothers Publishers.
Hann, Roy W., and Wesley P. James. 1974. *A Survey of the Economic and Environmental Aspects of an Offshore Deepwater Port at Galveston, Texas. Part II: Environmental Considerations.* Sea Grants Program. College Station: Texas A and M Unversity Press.
Harrowing, Frank Thomas. 1950. "The Galveston Storm of 1900." M.A. thesis, University of Houston.
Hayes, Charles W. 1879. *Galveston: History of the Island and the City.* 2 vols. Reprint, Austin: Jenkins Garrett Press, 1974.
Henry, W. K., D. M. Driscoll, and J. P. McCormack. 1975. *Hurricanes on the Texas Coast.* College Station: College of Applied Geosciences, Texas A and M University.
Hewitt, W. Phil. 1975. *The Czech Texans.* San Antonio: Institute of Texan Cultures, University of Texas. Reprint, 1998.

BIBLIOGRAPHY

———. 1994. *The Italian Texans*. San Antonio: Institute of Texan Cultures, University of Texas.

———. 1996. *The Jewish Texans*. San Antonio: Institute of Texan Cultures, University of Texas.

"The Housing Problem in Texas: A Study of Physical Conditions under Which the Other Half Lives." 1911. Reprinted from the *Galveston-Dallas News*, November-December 1911.

Houston, Matilda C. 1845. *Texas and the Gulf of Mexico or Yachting in the New World*. Philadelphia: G. B. Zieber and Co.

Hyman, Harold M. 1990. *Oleander Odyssey: The Kempners of Galveston, Texas*. College Station: Texas A and M University Press.

Jackson, Angela. 1998a. "Galveston's Sacred Heart Church Survives through Its Congregation." *Galveston Daily News*, 26 March, 9–10.

———. 1998b. "More Than a School." *Galveston Daily News*, 16 March, A1, A6.

Jordan, Terry G. 1966. *German Seed in Texas Soil: Immigrant Farmers in Nineteenth Century Texas*. Austin: University of Texas Press.

———. 1969a. "The German Settlement of Texas after 1865." *Southwestern Historical Quarterly* 73:193–212.

———. 1969b. "Population Origins in Texas, 1850." *Geographical Review* 59:83–103.

———. 1976. *Atlas of Texas*. Austin: Bureau of Business Research, University of Texas.

———. 1980. *Environment and Environmental Perceptions in Texas*. Boston: American Press.

———. 1981. "The 1887 Census of Texas Hispanic Population." *Aztlan, International Journal of Chicano Studies Research* 12:271–78.

———. 1982. "The Forgotten Texas State Census of 1887." *Southwestern Historical Quarterly*, 85:401–8.

———. 1983. "Immigration to Texas." In *Readings in Texas History*, ed. Cary D. Wintz, 93–130. Boston: American Press.

———. 1989. "Germans and Blacks in Texas." In *States of Progress: Germans and Blacks in America over 300 Years: Lectures from the Tricentennial of the Germantown Protest over Slavery*, 89–97. Philadelphia: German Society of Pensylvania.

Jordan, Terry G., with John L. Bean Jr. and William M. Holmes. 1984. *Texas: A Geography*. Boulder: Westview Press.

Jordan-Bychkov, Terry G. 2001. "The Creole Coast: Homeland to Substrate." In *Homelands: A Geography of Culture and Place across America*, ed. R. L. Nostrand and L. E. Estaville, 73–82. Baltimore: Johns Hopkins University Press.

Joseph, Samuel. 1969. *Jewish Immigrants to the United States from 1881 to 1910*. New York: Arno Press.

Kelly, Ruth Evelyn. 1975. "'Twixt Failure and Success: The Port of Galveston in the Nineteenth Century." M.A. thesis, University of Houston.

Kerr, Homer I. 1966. "Migration into Texas: 1860–1880." *Southwestern Historical Society Quarterly* 70:184–216.

BIBLIOGRAPHY

Kisch, Guido. 1949. "The Revolution of 1848 and the Jewish 'On to America' Movement." *Publications of the American Jewish History Society* 38:185–237.

Kozel, Gary, and Susanna French. 1999. "Preservation Frontline in the News: Galveston Historical Foundation Receives National Preservation Award" and "Galveston Historical Foundation's 'Operation Church Street.'" Washington, D.C.: National Trust for Historic Preservation. Web site: www.nthp.org

LaRoe, Edward T. 1979. "Barrier Islands and Their Management as Significant Ecosystems." In *Proceedings of the Gulf of Mexico Coastal Ecosystems Workshop*, ed. Paul L. Fore and Russell D. Peterson, 147–56. Albuquerque: U.S. Fish and Wildlife Service.

Larson, Erik. 1999. *Isaac's Storm*. New York: Crown Publishers.

Lembcke, Eva. 1985. *Passenger Lists for Galveston, 1850–1855*. Houston: Albert J. Blaha Sr., Publisher and Compiler.

Lowry, Jack. 2000. "The Storm of the Century." *Texas Monthly* 47:42–48, 50.

Macdonald, Keitha. n.d. *Galveston's 1900 Storm*. Galveston: Macdonald Publications.

Marinbach, Bernard. 1983. *Galveston: Ellis Island of the West*. Albany: State University of New York Press.

Marino, Samuel J. 1981. *Italian Newspapers in Texas*. Denton: Texas Women's University Press.

Mason, Herbert Molloy. 1972. *Death from the Sea*. New York: Dial Press.

Maury, Matthew F. 1849. "Great Commercial Advantages of the Gulf of Mex." *De Bow's Review* 7:510–23.

McComb, David. 1983. "The Houston-Galveston Rivalry." In *Houston: A Twentieth Century Urban Frontier*, ed. Francesco A. Rosales and Barry J. Kaplan, 18–32. Port Washington, N.Y.: Associated Faculty Press.

———. 1986. *Galveston: A History*. Austin: University of Texas Press.

McGowen, J. H., L. E. Garner, and B. H. Wilkinson. 1977. *The Gulf Shoreline of Texas: Processes, Characteristics, and Factors in Use*. Geological Circular 77-3. Austin: Bureau of Economic Geology.

McQuire, James Patrick. 1974. *The Greek Texians*. San Antonio: Institute of Texan Cultures, University of Texas.

———. 1976. *Stockfleth, Julius, Gulf Coast Marine and Lands Painter*. San Antonio: Trinity University Press; Galveston: Rosenberg Library.

———. 1993. *The Hungarian Texans*. San Antonio: Institute of Texan Cultures, University of Texas.

Miller, Ray. 1983. *Ray Miller's Galveston*. Austin: Cordovan Press.

Morrison, Andrew, ed. 1887. *The Industries of Galveston*. Galveston: Metropolitan Publishing Co.

———. 1890. *The Port of Galveston and the State of Texas*. Galveston: George W. Engelhardt.

Morse, S. F. B. 1902. *The Coast Country of Texas: One of the Most Promising Sections on the Line in the Great Southwest*. Houston: Press of Cummings and Sons, Printers.

Morton, Robert A., Orrin H. Pilkey Jr., Orrin H. Pilkey Sr., and William J. Neal. 1983. *Living with the Texas Shore.* Durham, N.C.: Duke University Press.

Mueller, Oswald F. Roemer, trans. 1935. *Texas with Particular Reference to German Immigration and the Physical Appearance of the Country.* San Antonio: Standard Printing Co.

Muir, Andrew Forest, ed. 1958. *Texas in 1837: An Anonymous Contemporary Narrative.* Austin: University of Texas Press.

Mullins, Marian Day, ed. 1959. *The First Census of Texas, 1829-36.* Washington, D.C.: National Genealogical Society.

Murdock, Steve H., et al. 1995. *Texas Challenged: Implications of Population Change for Public Service Demand in Texas.* College Station: Center for Demographic and Socioeconomic Research and Education, Texas A and M University.

National Fibers Information Center. 1987. *The Climates of Texas Counties.* Austin: Bureau of Business Research, Graduate School of Business, University of Texas.

Negro Population of the United States, 1790-1915. 1968. Washington, D.C.: U.S. Bureau of the Census, 1920. Reprint, Arno Press and *New York Times.*

Nesmith, Samuel P. 1975. *The Belgian Texans.* San Antonio: Institute for Texan Cultures.

Newbolt, Lawrence E., and John B. Herbich. 1979. "Hydrology of Coastal Waters." In *Proceedings of the Gulf of Mexico Coastal Ecosystems Workshop,* ed. Paul L. Fore and Russell D. Peterson. Albuquerque: U.S. Fish and Wildlife Service.

Newcomb, W. W., Jr. 1961. *The Indians of Texas: From Prehistoric to Modern Times.* Austin: University of Texas Press.

Ornish, Natalie. 1989. *Pioneering Jewish Texans.* Dallas: Texas Heritage Press.

Ousley, Clarence. 1900. *Galveston in Nineteen Hundred.* Atlanta: William C. Chase.

Page, Frederic Benjamin. 1984. *Prairiedom: Rambles and Scrambles in Texas.* New York: Paine and Burgess.

Paine, Jeffrey G., and Robert A. Morton. 1986. *Historical Shoreline Changes in Trinity, Galveston, West, and East Bays, Texas Gulf Coast.* Geological Circular 8603. Austin: Bureau of Economic Geology.

Panitz, Esther L. 1965. "In Defense of the Jewish Immigrant, 1891-1924." *American Jewish Historical Quarterly* 55. Reprinted in *The Jewish Experience in America,* ed. Abraham J. Karp. New York: Ktav Publishing House, 1969.

Pearson, Charles E. 1994. *Cultural Resources Remote-Sensing Survey, Offshore Borrow Areas for Beach Replenishment Project, 10th Street to 103rd Street, Galveston County, Texas.* Report prepared for the City of Galveston. Baton Rouge: Coastal Environments, Inc.

Perkes, Kim Sue Lia. 1998. "Shalom Y'all: Musing on Joys, Trials of Texas Jews." *Austin-American Statesman,* 13 December, A1, A14.

Pratt, Willis W., ed. 1954. *Galveston Island (or) A Few Months off the Coast of Texas: The Journal of Francis C. Sheridan, 1839-1840.* Austin: University of Texas Press.

Price, Granville. 1930. "A Sociological Study of a Segregated District." Thesis, University of Texas, Austin.

BIBLIOGRAPHY

Rabelais, Nancy N. 1979. "Ecological Values of Selected Coastal Habitats." In *Proceedings of the Gulf of Mexico Coastal Ecosystem Workshop,*. ed. Paul L. Fore and Russell D. Peterson, 191-203. Albuquerque: U.S. Fish and Wildlife Service.

Reese, James V. 1968. "Early History of Labor Organizations in Texas, 1838-1876." *Southwestern Historical Quarterly* 72:1-20.

———. 1971. "The Evolution of an Early Texas Union: The Screwman's Benevolent Association of Galveston, 1866-1891," *Southwestern Historical Quarterly* 75:158-85.

Remmers, Mary W. 1997. *Going down the Line: Galveston's Red-Light District Remembered.* Galveston: By the author.

Rice, Bradley Robert. 1977. *Progressive Cities: The Commission Movement in America, 1901- 1920.* Austin: University of Texas Press.

Rice, Lawrence D. 1971. *The Negro in Texas, 1874-1900.* Baton Rouge: Louisiana State University Press.

Richardson, David. 1862. *The Texas Almanac for 1862 with Statistics, Historical, and Miscellaneous Sketches Relating to Texas.* Houston: n.p.

Ricklis, Robert A. 1996. *The Karankawa Indians of Texas: An Ecological Study of Cultural Tradition and Change.* Austin: University of Texas Press.

Rosales, Francisco A., and Barry J. Kaplan. 1983. *Houston: A Twentieth Century Urban Frontier.* Port Washington, N.Y.: Associated Faculty Press.

Sacred Places. n.d. 3 vols. Galveston: Galveston Historical Foundation.

Sance, Melvin M., Jr. 1975. *The Afro-American Texans.* San Antonio: Institute of Texan Cultures, University of Texas.

San Giovanni Italian-American Association of Galveston. 1992. "Honoring the Italian Vault, 1888." Pamphlet of a talk given by the Most Reverend Joseph A. Fiorenza, bishop of the Diocese of Galveston-Houston, at the 11 October 1992 ceremony at the vault in conjunction with the quincentennial celebration of Christopher Columbus's discovery of the Americas.

Scheibe, Merri Jane. 1992. "Galveston's First Re-construction: 1865-1874." Thesis, University of Houston, Clear Lake.

Shaw, Tom. 1999. "Just around the Corner: Washington, DC Exhibit Showcases Isle 'Mom and Pop' Stores of Yesteryear." *Daily News,* 25 November, A1-A12.

Shpall, Leo. 1945. "Spreading the Jewish Migrant in America: The Galveston Experiment." *Jewish Forum* 28:6-8,119-20, 139-40, 144, 156-58.

Shuman, Bernard. 1969. *A History of the Sioux City Jewish Community, 1869 to 1969.* Sioux City, Iowa: Bolstein Creative Printers.

Sibley, Marilyn McAdam. 1968. *The Port of Houston: A History.* Austin: University of Texas Press.

Silverthorne, Elizabeth. 1982. *Ashbel Smith of Texas: Pioneer, Patriot, Statesman, 1805-1886.* College Station: Texas A and M University Press.

Simon, Maurice, ed. 1937. *Speeches, Articles, and Letters of Israel Zangwill.* London: Soncino Press.

Simonds, James Persons. 1913. *Report of a Sanitary Survey of the City of Galveston.* Galveston: Galveston Commercial Association.

BIBLIOGRAPHY

Sisters of Charity of the Incarnate Word. 1996. *A Pattern of Love: The story of St. Mary's Hospital, in Words and Pictures, Commemorating 129 Years of Service to the Community of Galveston, 1867-1966.* Houston: Sisters of Charity of the Incarnate Word Health Care System.

Sketches and Views In and About Greater Galveston: Its Unique History, Commanding Location, Enormous Commerce, Leading Business Houses, Delightful Beach, Great Sea Wall, Beautiful Homes, Churches, Schools, Parks, Drives, etc. 1906. Souvenir edition of *Galveston Tribune*.

Southern Pacific Railroad, Freight Traffic Division. 1902. *Largest Pier in the World: Southern Pacific Freight Terminals, Galveston, Texas.* Galveston: Southern Pacific Railroad.

Stanley, Donald W. 1992. *Water Quality and Fisheries: Galveston Bay, Historical Trends.* Washington, D.C.: U.S. Department of Commerce, National Oceanic and Atmospheric Administration.

Steinert, Wilhelm. 1999. *North America, Particularly Texas in the Year 1849: A Travel Account.* Trans. Gilbert J. Jordan and ed. T. G. Jordan-Bychkov. Dallas: DeGolyer Library and William P. Clements Center for Southwest Studies, Southern Methodist University.

Stephens, A. Ray, and William M. Holmes. 1984. *Historical Atlas of Texas.* Norman: University of Oklahoma Press.

Storm Forum. Web site of the *Galveston Daily News*. Submission by De Voti Family, 2000. 3 pp.

Struve, Walter. 1996. *Germans and Texans: Commerce, Migration, and Culture.* Austin: University of Texas Press.

Swanton, J. R. 1952. *The Indian Tribes of North America.* Bulletin 145. Washington, D.C.: Bureau of Indian Ethnology.

Szajkowski, Zosa. 1967. "Paul Nathan, Lucien Wolf, Jacob H. Schiff and the Jewish Revolutionary Movements in Eastern Europe, 1903-1917." *Jewish Social Studies* 24:1-14.

Tremblay, Thomas A. 1992. "A GIS Study of Cumulative Wetland and Habitat Change: The Virginia Point Quadrangle, Galveston County, Texas." M.A. thesis, University of Texas, Austin.

Trillin, Calvin. 1996. *Messages from My Father.* New York: Noonday Press, Farrar, Strauss and Giroux.

Turner, Elizabeth Hayes. 1997. *Women, Culture, and Community: Religion and Reform in Galveston, 1880-1920.* New York: Oxford University Press.

Tyler, Ron, and Lawrence R. Murphy, eds. 1974. *The Slave Narratives of Texas.* Austin: University of Texas Press.

Wade, Richard C. 1964. *Slavery in the Cities: The South 1820-1860.* Oxford: Oxford University Press.

Walden, Don. 1990. "Raising Galveston." *American Heritage of Technology and Invention* (winter): 8-18.

BIBLIOGRAPHY

Waldman, Norris D. 1928. "The Galveston Movement: Another Chapter from the Book Which May Never Be Written." *Jewish Social Science Quarterly* 43:197–205.

Wall, E. L., ed. 1928. *The Port Situation in Galveston.* Galveston: Galveston News Company.

Walton, Anne H., and Albert W. Green. 1993. *Probable Causes of Trends in Selected Living Resources in the Galveston Bay System.* Report 33. Galveston: Galveston Bay National Estuary Program.

Ward, George H. 1993. *Dredge and Fill Activities in Galveston Bay.* Report 28. Galveston: Galveston Bay National Estuary Program.

Webb, Walter Prescott, ed. 1952. *The Handbook of Texas.* Austin: Texas State Historical Association.

Weber, David J. 1982. *The Mexican Frontier, 1821–1846: The American Southwest under Mexico.* Albuquerque: University of New Mexico Press.

Weems, John Edward. 1900. *A Weekend in September.* New York: Holt.

Wheeler, Kenneth W. 1968. *To Wear a City's Crown: The Beginnings of Urban Growth in Texas, 1836–1865.* Cambridge: Harvard University Press.

White, Edna McDaniel. 1965. *East Texas Riverboat Era and Its Decline.* Beaumont: Labelle Printing and Engraving Co.

White, Gifford, ed. 1966. *1840 Census of the Rupublic of Texas.* Austin: Pemberton Press.

Wohlgelernter, Maurice. 1964. *Israel Zangwill: A Study.* New York: Columbia University Press.

Woodward, Earl F. 1972. "Internal Improvements in Texas in the Early 1850s." *Southwestern Historical Quarterly* 76:161–82.

Woolfolk, George R. 1976. *The Free Negro in Texas, 1800–1860: A Study in Cultural Compromise.* Ann Arbor: University of Michigan Press.

Wygant, Larry. 1992. "Medicine and Public Health in Galveston, Texas: The First Century." Ph.D. diss., University of Texas Medical Branch of Galveston.

Young, Earle B. 1997. *Galveston and the Great West.* College Station: Texas A and M University Press.

Index

Numbers in *italics* denote illustrations; those in **boldface** denote tables.

Addison, G. M., 26
African American(s): and city ordinances, 48–49, 57; after Civil War, 79; and desegregation, 146; in early Galveston, 22, 42, 46–49; in 1850 and 1860 census, 42, 46, **46**; and Emancipation Proclamation, 84–85, *86;* and Great Storm, 101, 103; and housing, 46, 69, 79, 145–46; and labor unions, 87–88; occupations of, 47, 49, *55;* during Reconstruction, 85–86; and segregation, 79, 80–81, 83, 110, 142, *144;* as slaves, 47–48, 52; in social structure, 57; under Spanish and Mexican rule, 47; and tourism industry, 150. *See also* churches; schools; social and service organizations
Allen, Joseph M., 33
Alleys and Back Buildings of Galveston, The (Beasley), 10
architecture, 3, 50, *51,* 74
Ashton Villa, 5
Audubon, John James, 29–30
Aury, Louis-Michel, 20–21, 47
Austin, Stephen F., 23, 29

Bandera (Texas), 77
Barr, Alwyn, 79
barrier islands, 6
Beasley, Ellen, 10; on alley houses, 46, 79; on architectural preservation, 147
Beaumont (Texas), 12, 62, 143
Belfiglio, Valentine J., 106

Belgians: during Civil War, 68; in 1880 census, 79; in 1900 census, **94**
Belo, A. H., 73
Bi-Racial Committee on Desegregation, 146
Bishop's Palace, 74, *75*
Bleinke, Joseph, 42
Bolivar Peninsula, 116
Bollaert, William, 24
Borden, Gail, 33, 48
born on the island (BOI), 12, 143
Bowie, Jim, 47
Brazos River, *17,* 62, 107
Bremen, German port of, 31, 70, 75, 77; and Galveston Movement, 128; pre-departure information in, 35; U.S. destinations of ships from, **34**
Broadway, 20, 83; and Great Storm, 99; as highest point, 113; historic structures on, 5, *75;* 148–50
Brown, Major John Henry, 71
Buffalo Bayou, 56, 70; deep-water channel, 94, 141; railroad, 63
Burnet, David, 30

Cabeza de Vaca, Alvar Nunez, 15, 17–20, *19*
Campbell, Albert, 101
Canadians: in 1850 census, 41; in 1880 census, 79; in 1900 census, **94**
Cantini, Arminio, 105
Caribbean, 21; region, 11, 27, *152*

INDEX

Cassel, 129, 131
Castro, Henri, 37
cemeteries: creation of, 32–33; Italian Vault, 107–8; Jewish, 125–26
census: manuscript, 29; contents of, 7; of 1850, 35, 37, 41–42, 46, **46**; of 1860, **46**; of 1880, 78–79
Center for American History, University of Texas, 8
Chinese: businesses, 108, *108*; residential patterns, *104*
churches, 5, 31, *53*; African Baptist, 52; African Methodist Episcopal, 52, 83–85, *86*; German Evangelical Lutheran, 42; and Great Storm, 101, *102*; Methodist, 83–84; Presbyterian, 107; Sacred Heart Catholic, 99, 101, 107, *107*; St. Joseph's Catholic, 42, *45*; St. Mary's Catholic, 51, 107; St. Patrick's Catholic, 101, 107, *115*
Civil War, 39–41, 67–69
Clayton, Nicholas, 74
Cline, Isaac, 99
Cohen, Rabbi Henry, 127; and Galveston Movement, 119, 121, 129
Colorado River, *17*, 26, 63
Comal County (Texas), 40–41
Cook, Margaret Devoti, 93
Corpus Christi (Texas), 16
cotton: compresses, 87, 90; export of, 54, 70, 93; at Galveston Wharf, *55*; production, 61; transportation of, 62, 63, 87
Creole Coast, 9; characteristics of, 10–11, 150–51
Cuney, Norris Wright, 72, 87, 88
customs house, 2, *3*; Mexican, 23, 29; records, 27, 30; republic of Texas, 32
Czechs, 76, 78; in 1900 census, *94*

Danes: during Civil War, 68; in 1850 census, 41; in 1880 census, 79; in 1900 census, **94**; residential patterns, *81*
data sources, 7–9
Devoti, Joseph and Margaret, 92–93
Dickinson (Texas), 107
Dirks, Friedrich (Ernst), 37
diseases, 24, 56; introduced by Europeans, 18. *See also* yellow fever

Dominican, 83
Dowell, Dr. Greensville, 79
Dutch, 15; during Civil War, 68; in 1900 census, **94**
Dyer, Isidore, 125

East End Historic District, 147
egalitarian myth, 10
Eisenhour, Virginia, 48
Ellenburger, Joseph, 28–29, 31
Elisa, 147
English, 16; during Civil War, 68; in 1850 census, 41; in 1880 census, 79; in 1900 census, **94**; labor union members, 87; residential patterns, *81*; social organizations, 94
Esteban, 19–20, 42
explorers, 15–23, *19*

Ferber, Edna, 1
Fiddler on the Roof, 120
fieldwork, 2–3
Finn, Tim, 72
Fisher, Charles, 30
Flake, Ferdinand, 37, 39–40
Flora, 28–29, 31
Fornell, Earl, 54, 56, 141
Fort Crockett Military Reservation, 115
Fort Point, 30, 67
Fort San Jacinto Reservation, 115
French: during Civil War, 68; in 1850 census, 41; in 1880 census, 79; explorers, 17, *19*, 20; in 1900 census, **94**; settlement, 20

Galveston: as capital of Texas, 30; Caribbean connection, 11, *152*; during Civil War, 67–69; during Depression, 145; 1870s description of, 60; fishing industry, 145; as Gulf Coast pivot city, 9, 61, *151*, 152–53; nicknames for, 1, 33–34, 60–61, 140; origin of name, 20; perceptions of, 1, 4, 6–7, 23–27, 29–30, 36–37; population growth of, 6; Sandusky map of, 13–14, *73*; tourism industry, 145, 148, 150
Galveston, city of: architecture, 3, 50, *51*, 74; amenities in, 49–50, 73; as cosmo-

INDEX

politan, 3, 34, 61, 75, 93; economy, 89; in 1871, *73;* foreign consulates in, 35, 94; as global city, 27, 35, *53,* 93–94; government, 109–12; grade-raising project, 116–18, *116, 117;* Great Storm impact on, 99–103; *100, 102;* growth of, 30; incorporation of, 4; international trade, 54, 60; ordinances, 48–49; population in U.S. census, 35, 72–73, 91, 93, 140; preservation efforts, 147–50, *149;* religion in, 50–52; seawall construction, 113–15; social stratification in, 12, 57, 90–91, 141–43, 150; state firsts in, 34, 64, 73, 125; survey of, 32–33; voting laws in, 57. *See also* churches; schools; social and service organizations; transportation

Galveston, port of: and cotton, 90, 93; description of entering, 34; elite control of, 141; records, 8; regional importance of, 89; rivalry with Houston, 56, 62–65, 89, 94–95, 141, 145; study of, 145; Union embargo on, 68

Galveston Bay, 23, 63; causeway bridge, 64; and hurricanes, 95; Spanish in, 20; storm surge in, 113–14; transportation in, 71

Galveston City Company, 32–33, 52, 56

Galveston Civilian, 39

Galveston Community Book, 75

Galveston Cotton Exchange, 73, 90

Galveston County: population, 47; seat, 33

Galveston Daily News, 61, 90; Emancipation Proclamation reports, 79; on Great Storm, 92–93, 101; hurricane warnings, 91; on migration, 72; port study, 145; on yellow fever, 72

Galveston Deep Water Commission, 110–11

"Galveston: Gateway to the Texas Empire," 138–39

Galveston and the Great West (Young), 10

Galveston Harbor: perceptions of, 23, 29; ships entering, **8,** 27, 30; U.S. immigration center, 76; improvements, 71, 89, 118; wharfs in, 54, 56, 88

Galveston Historical Foundation (GHF), 147–48, 15

Galveston Historical Society, 54, 147

Galveston Island: as barrier island, 6, 25; climate of, 24; drinking water, 25–26, 54; early names for, 19–20; flora and fauna of, 16, 24; human alteration of, 6, 23, 56, 113, *114;* and hurricanes, 95; location of, *2,* 12; military bases on, 21, 30, 32, 67, 115; plantations on, 48; physical environment of, 6, 18, 50; as prison, 30; and storm surges, 97

Galveston Medical Journal, 79

Galveston Movement, 119–37, *130;* destinations, 120, 131–32, *133;* participants' occupations, 129, 131; transportation, 124–25, 128–29. *See also* Jews

Galveston Plan, 111–12

Galveston Tri-Weekly Telegraph, 68

Galveston Wharf and Cotton Press Company, 54, 89–90

Galveston Weekly News: commentary on migration, 74; report of German arrivals, 28

Galveston Zeitung, 34

Galvez, Bernardo de, 20

Garten Verein, 39, *42*

Geifman, Gershom, 119

Geographic Information System (GIS), 9; maps, *40, 43–44, 53, 80–82, 102, 104–5, 108*

German(s): businesses, 39, *40;* churches, 31, 42, *45;* during Civil War, 39–41, 68; descendants, 9; in 1880 census, 78; food preferences, 37; impact on cultural landscape, 31, 37, 54; labor union members, 87; newspaper, 34; in 1900 census, **94;** occupations, 28, 37, **41;** population, 3, 31; residential patterns, *43;* schools, 31; settlement in Texas, 35–41; on slavery, 39–41; and social organizations, 52, 54, 94; social status of, 39, 57

German Seed in Texas Soil (Jordon), 9

Goliad (Texas), 21–22

Grand Opera House, 93

Great Storm, 98–99; changes in aftermath of, 109–18; effects on city, *100,* 101–3, *102, 103;* effects on immigrants, 100–101, 103–5; memories of, 92–93

INDEX

Greeks: in 1880 census, 79; and fishing, 24; residential patterns, *80, 104*
Green, Casey Edward, and Shelly H. Kelly, 10
Gregory, General Edgar, 84–85
Grijalva, Juan de, 18
Groesbeck, John, 32
Gudzikowski, Laurie, 131
Gulf Coast, 91; characteristics of cities, 11; hurricanes, 96, 118; pivot city of, 9, 61, *151*, 152–53; ports, 89; region, 9–10; and yellow fever, 27, 125. *See also* Texas Gulf Coast
Gulf Coast Fishermen's Association, 145
Gunther, John, 141

Haitians, 15, 20–21
Hamilton, Jeff, 47
Hayes, Elizabeth Turner, 10, 27
Heiress, 59
Hendrick, Layne, 15
Hill Country (Texas), 35, 37
housing, 3–4; for African Americans, 46, 69, 79, 145–46
Houston, 3, 46, 140; African Americans in, 46, 48; during Civil War, 69; port rivalry with Galveston, 56, 89, 94–95, 141, 145; transportation rivalry with Galveston, 62–65, 70–71, 142
Houstoun, Matilda, 24
Hungarians, 76, 78; during Civil War, 68; in 1900 census, **94**
hurricanes, 3, 6; in 1839, 34; in 1867, 72; in 1875, 91; historic storm events, 95–96; in 1909 and 1915, 118; perceptions of danger, 24, 96–97; surge heights, *97. See also* Great Storm

immigrants: and city government, 112; during Civil War, 68; impact of, 4–5, 75; and labor unions, 86–88; in 1900 census, **94**; occupations of, 54, *55*; residential patterns, *82. See also specific nationalities*
immigration laws, 76, 140; and Galveston Movement, 135–37

Index of Naturalization Records in Texas Courts, 8
Indianola (Texas), 64, 77; and hurricanes, 91, 109, 118
Institute of Texan Cultures, 131
Irish, 74–75; in 1850 census, 41; in 1880 census, 79; labor union members, 87; in 1900 census, **94**; residential patterns, *81;* social organizations, 74–75, 94
Isaac's Storm (Larson), 91
Italian(s): businesses, 105, 108, *108,* 145, *146;* during Civil War, 68; in 1850 census, 41; in 1880 census, 79; descendants, 9; fishing, 24; and Great Storm, 93, 103–5; immigrants, 76, 105–7; newspapers, 108–9, *109;* in 1900 census, **94**; residential patterns, *80, 104, 105;* social organizations, 94, 108–9

Jamaica Beach, 16
Jewish Immigrants' Information Bureau (JIIB), 128–29
Jewish Territorial Organization (JTO), 119, 123–24
Jews: cemeteries for, 125–26; in city politics, 125; first congregation of, 52; German, 35–36, 121; introduce oleanders, 125, *126;* Russian, 119–25, 128–37, *130;* social organizations, 125–28. *See also* Galveston Movement; synagogues
Jordan, Terry: on Creole Coast, 9–11, 150–51; on Germans and slavery, 39–41; on Germans in Texas, 9, 37; on hurricanes, 97; on origin of German settlers, *38;* settlers' misconceptions, 23–24

Kansas City, 125, 131–32, *133*, 134
Karankawa Indians: contact with Europeans, 17, *18;* coping with climate, 25; hunting parties of, 15–16; settlements of, 16, *17;* U.S. troops' killing of, 22
Kelly, Ruth Evelyn, 89
Kempner, Harris, 72, 126
A Kind of Magic (Ferber), 1
Kirwin, Father James, 101
Kopperl, Mortiz, 78

INDEX

labor unions, 86–88
Lafitte, Jean: control of Galveston Island, 21–23; and slaves, 47
Lafitte, Pierre, 21–22
Lamar, Mirbeau B., 26
Larson, Erik, 91, 99
La Salle, Robert Cavelier sieur de, 20
Liberty (Texas), 22
Long, Dr. James, 22
Los Rojos de Roma, 138–39
Louisiana, 20–21, 26; and Great Storm, 96
Lowry, Jack, 99

Macdonald, Kietha, 99
Madera, Frank, 101
Madison, Mary, 47
Magruder, General John, 68–69, 71
Maillard, N. Doran, 24
Mallory Line, 70–71; and Galveston Wharf Company, 90; and labor unions, 88
Maniscalco, Pietro, 107
Matagorda (Texas), 16–17, 63
Maury, Matthew, 24
McKinney, Thomas F., 32
Melville, Herman, 93
Menard, Michael, 31–32; and Galveston Wharf and Cotton Press Company, 54; house of, 50, *51*; as slave owner, 48
Messina, Maria, 107
methods of research, 3, 7–9
Mexicans: as early settlers, 15, 23; in 1850 census, 41; in 1880 census, 79; in 1900 census, **94**; residential patterns, *80*; in social structure, 58
Mina, Francisco, 20–21
Moby Dick (Melville), 93
Moczygemba, Rev. Leopold, 77
Moody: convention center site, 73; gardens, 150
Morgan Line, 33, 70–71, 88
Murney, Will, 100

Nacogdoches (Texas), 20, 31, 127
native-born elite, 57, 87, 110, 141; and Galveston City Company, 56–57; and Galveston Wharf Company, 90

New Braunfels (Texas), 35
New England, 71; teachers from, 52, 82
New Orleans, 11, 20–21, 51; and Galveston Movement, 124; and immigration, 88; newspaper report, 58; port of, 89; transportation connections, 27, 30–31, 33, *34*, 54, 60, 63, 69, 75, 107; and yellow fever, 26, 125
News Orleans Picayune, 58
newspapers, 33, 39, *53*; German language, 34, 37, 39; Italian language, 108–9, *109*. See also *Galveston Daily News*; *Galveston Weekly News*
New York Herald, 61
Nicolini, Clemente, 108
North German Lloyd Line, 124, 129
Norwegians: in 1900 census, **94**; residential patterns, *81*

Odin, Father John, 51
Old Central Cultural Center, 83, *84*
oleanders, 3, 50, *126*; visitors' descriptions of, 7, 25
Ornish, Natalie, 122–23
Osterman, Joseph, 33
Osterman, Rosanna, 125

Page, Fredrick Benjamin, 4
Paine, Jeffrey G., and Robert A. Morton, 95
Panna Maria (Texas), 77
Pelican Island, 76, *76*, 88, 94
Perry, Colonel Henry, 21
Pineda, Alonso Alvarez de, 18
Pioneer, 70
Point Bolivar, 72–73
Polish: in 1850 census, 41; immigrants, 76–78; in 1900 census, **94**
Port Arthur (Texas), 11–12, 143
Portuguese: during Civil War, 68; in 1880 census, 79
Powers, Bill and Christine, 59–60
Prairiedom, 31

railroads. *See* transportation
Ramirez, Betto, and Grace Arjona-Ramirez, 138–39

INDEX

Reese, James V., 87
residential patterns, 69, 143; analysis of, 7; in 1857, *43*, *44;* in 1880, 79–81, *80–82;* and Great Storm, 103, *104*, *105*
Richardson, William, 63
Richardson's News, 63
Rizzo, Lucy, 104
Roemer, Dr. Ferdinand, 34
Rosenberg Library, 7, 13–14, 83, **84;** customs house records, 27; microfilm records, 30; scholastic census, 83
Russians, 9; in 1900 census, **94**. *See also* Jews

Saccarap, 48
St. Mary's Hospital, 99
St. Mary's Orphanage, 100–101, *102*, 103–4; site of, 115
San Antonio (Texas), 37, 63; African Americans in, 86; immigrant settlement near, 77–78; population size of, 93
Sandusky, William, 77
Sandusky map, 13, 20, *73*
Santa Ana, 30
Scandinavian: residential patterns, *81;* labor union members, 87. *See also specific nationalities*
Schiff, Jacob, and Galveston Movement, 121, 123–25, 135–36
schools, 52; African American, 82, 85, 103; Ball High School, 83, **84;** Catholic University of St. Mary's, 52; Central High School, 83, **84,** 85; desegregation of, 146; East District School, 103; Galveston Female Collegiate Institution, 52; Hebrew Orthodox, 128; in 1900 census, *103;* Ursuline Academy for Girls, 51–52
Scottish: in 1850 census, 41; in 1880 census, 79; labor union members, 87; in 1900 census, **94;** residential patterns, *81*
Sealy Mansion, 150
seawall, 113–15, *114*
Seeligson, Michael, 125
segregation indexes, 9, 143, *144*
Seguin, Juan, 32
Sheridan, Francis, 23–24, 33, 47

ship passenger lists, 8, 29
Sigler, Henry, 47
Smith, Dr. Ashbel, 26–27, 30
social and service organizations, 52, 54, 94; African American, 58, 81; Irish, 74–75; Italian, 108–9; Jewish, 125–28
Spanish, 15; during Civil War, 68; in 1850 census, 41; in 1880 census, 79; explorers, 18–20; in 1900 census, **94;** residential patterns, *80*. *See also* Cabeza de Vaca
steamship service. *See* transportation
Steinert, Wilhelm, 25, 31, 34
Strand, 65, 144, 148; in 1869 fire, 72; historic district, 2, *3*, 147; immigration station, 88
Swedish: during Civil War, 68; in 1850 census, 41; in 1880 census, 79; in 1900 census, **94;** residential patterns, *81*
Sweet, Alexander E., 95
Swiss: during Civil War, 68; in 1850 census, 41; in 1880 census, 79; in 1900 census, **94**
Sydnor, John, 47
synagogues: Congregation Ahavas Israel, 128; Congregation Beth Jacob, 125; Temple B'nai Israel, 125, *127*, 128

Texas: African Americans in, 42, **46,** 85–86, 142; commercial treaties with, 54; Constitution, 82; Czech settlement in, 78; first African settler in, 18–19; German settlement in, 35–41; immigrants' perception of, 6; Italians in, *106, 109;* legislature, 63, 110–11; Polish settlement in, 77–78; during Reconstruction, 85; slavery in, 47; transportation in, 61–65; voting laws in, 57; war of independence, 30
Texas Gulf Coast, 4, 23; German settlement along, 39; hurricanes, 91, 95–96, 109, 118; Karankawa settlement along, 16–18, *17;* transportation, 62, 67
Texas Heroes Monument, *149*, 150
Texas Seaport Museum, 8, *148*
Texas State Archives, 8
Third Coast, 61, 150–51

INDEX

Three Trees, 22–23
Through a Night of Horrors: Voices from the 1900 Great Storm (Green and Kelly), 10
tourism industry, 145, 148, 150
transportation: Galveston-Houston rivalry, 62–65, 70–71, 142; interurban railroads, 66, *67;* railroads, 62–66, 65; rivalry with Houston, 62–65, 70–71, 142; steamship service, 30, 54, 69–71, 124; streetcars, 65–66, *66. See also* Mallory Line; Morgan Line; North German Lloyd Line
Trillin, Calvin, 120, 131–32
Trinity River, 22, 26, 31, 62
tropical storms, 6, *96*
Truehart, Henry M., 74
Turner, Frederick Jackson, 47
Two Years in Mexico, or The Emigrant's Guide (Brown), 71

U.S. Census of Population Manuscript Census, contents of, 7. *See also* census, manuscript
U.S. Immigration and Naturalization Service, 8
University of Texas Medical Center: opening of, 91, 93; property, 88; site of, 21
Ursuline Convent, 101

Vidor, Charles, 78

Waldman, Morris, 119, 128–29
Welsh: in 1850 census, 41; in 1880 census, 79; in 1900 census, **94;** residential patterns, *81*
Wheeler, Kenneth W., 25
Williams, Samuel May: and Galveston City Company, 32; and Galveston Wharf and Cotton Press Company, 54; home of, 50, 147; as politician, 57
Women, Culture and Community: Religion and Reform in Galveston, 1880–1920 (Hayes), 10

yellow fever: epidemic of 1840, 34; epidemic of 1867, 72, 126; impact on trade, 64; threat of, 24, 27, 125
Young, Earle, 10

Zangwill, Israel, 121, 121–25, 134–36

ABOUT THE AUTHOR

SUSAN WILEY HARDWICK is an associate professor of geography at the University of Oregon and formerly a professor of geography and associate director of the Grosvenor Center for Geographic Education at Southwest Texas State University. As a human geographer she has specialized in the settlement of the American West, especially the urban West. Dr. Hardwick is best known for her book *Russian Refuge: Religion, Migration, and Settlement on the North American Pacific Rim* (1993) and her coauthored university textbook *Geography for Educators: Standards, Themes, and Concepts* (1996). In 1995 she was awarded the statewide California State University Outstanding Professor Award.

OTHER GEOGRAPHICAL BOOKS IN THE SERIES

*The American Backwoods Frontier:
An Ethnic and Ecological Interpretation*
TERRY G. JORDAN AND MATTI KAUPS

Cities and Buildings: Skyscrapers, Skid Rows, and Suburbs
LARRY R. FORD

The Cotton Plantation South since the Civil War
CHARLES S. AIKEN

Delta Sugar: Louisiana's Vanishing Plantations
JOHN B. REHDER

Let the Cowboy Ride: Cattle Ranching in the American West
PAUL F. STARRS

The Los Angeles River: Its Life, Death, and Possible Rebirth
BLAKE GUMPRECHT

*Manufacturing Montreal:
The Making of an Industrial Landscape, 1850 to 1930*
ROBERT LEWIS

*Measure of Emptiness:
Grain Elevators in the American Landscape*
FRANK GOHLKE, WITH A CONCLUDING ESSAY BY JOHN C. HUDSON

The Mountain West: Interpreting the Folk Landscape
TERRY G. JORDAN, JON T. KILPINEN, AND CHARLES F. GRITZNER

The New England Village
JOSEPH S. WOOD, WITH A CONTRIBUTION
BY MICHAEL P. STEINITZ

The North American Railroad: Its Origin, Evolution, and Geography
JAMES E. VANCE, JR.

*The Pennsylvania Barn:
Its Origins, Evolution, and Distribution in North America*
ROBERT F. ENSMINGER

*Pride in the Jungle:
Community and Everyday Life in Back of the Yards Chicago*
THOMAS L. JABLONSKY

*The Promise of Paradise:
Recreational and Retirement Communities in the
United States since 1950*
HUBERT B. STROUD

*The Spanish-American Homeland:
Four Centuries in New Mexico's Río Arriba*
ALVAR W. CARLSON

To Build in a New Land: Ethnic Landscapes in North America
EDITED BY ALLEN G. NOBLE

Unplanned Suburbs: Toronto's American Tragedy, 1900 to 1950
RICHARD HARRIS